Draggermen

Draggermen

FISHING ON GEORGES BANK

by George Matteson

Four Winds Press New York

To Rath

LIBRARY OF CONGRESS CATALOGING IN PUBLICATION DATA

Matteson, George.
  Draggermen.
  SUMMARY: Follows the activities on board a
trawler as it searches for yellowtail flounder, cod,
and haddock off the New England coast.
  1. Fisheries—Georges Bank—Juvenile literature.
2. Trawls and trawling—Juvenile literature.
3. Fishermen—Juvenile literature. [1. Trawls
and trawling. 2. Fishermen] I. Title.
SH331.15.M37   639'.22'09744   78–21767
ISBN 0–590–07534–9

Published by Four Winds Press
A division of Scholastic Magazines, Inc., New York, N.Y.
Copyright © 1979 by George Matteson
All rights reserved
Printed in the United States of America
Library of Congress Catalog Card Number: 78-21767
1 2 3 4 5 83 82 81 80 79

"Every boat operates a little different," says the captain, "and every day's a little different too. If you want to learn about fishing you got to go on a lotta different boats. You got to work under people who do things right and you got to watch carefully to see exactly what they do. So when you got your own boat you will know when you are doing different from them, and you will know when you are doing better."

The New England Fishery has evolved through hundreds of years and the combined intelligence of thousands of lifetimes. The craft is life to life a little changed, a little marked, like a tool gone hand to hand. It is a tough business to learn, but the first hour at sea is a start, and even a day spent in careful observation will net a plentitude of knowledge.

I was fortunate to have had many days at sea on several different vessels for which, in addition to the principals of this book, I deeply thank Capt. Jim McCauley and crew of the trawler *Alliance*, Capt. Frank Lami and crew of the dragger *Jerry & Jimmy*, Dr. James E. Hankes of the National Marine Fisheries Service, and Robert Parisi of Parisi Seafood Company, New Bedford, Massachusetts. Each of these men gave freely of his valuable time and knowledge.

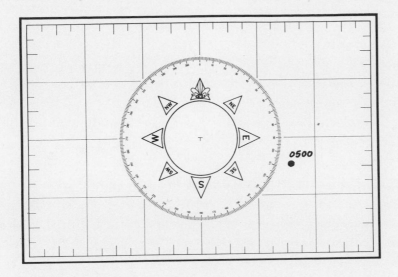

0500

The sun will rise in fifty-five minutes.

It is Sunday, September 4th. From latitude 41° 22′ north, longitude 66° 44′ west, from what is called the Southeast Part of Georges Bank, it is 145 miles to the island of Nantucket, the coast of Massachusetts, the nearest land. Sunrise will not occur on Nantucket until thirteen minutes after Southeast Part.

Since just after midnight the fishing vessel, *Elise G.*, has lain motionless in the North Atlantic. All night the wind has been calm. A low ocean swell has rolled in from the south. Among these easy waves the vessel has rocked gently and the crew, in the 4½ hours that it sleeps this night, sleeps well.

The ship's engine is shut down. The only sound from the engine room is the constant yammering of one of the ship's two electrical generators. One or the other of these identical machines has been running virtually nonstop since the vessel was launched. The crew are so accustomed to the racket that they would wake up in alarm if it were silenced.

There are four men; three asleep and one awake on watch in the wheelhouse, the place from which the vessel is controlled. Each man has stood a 1½-hour watch tonight. First John Aksdal, then Rob Rule, then Tom Montgomery. Now, from 4:30, the captain, Kaare Gjertsen, stands his watch until it is time to begin fishing at dawn.

The night has been moonless and there has been a thick fog. Not even starlight has been able to reach the sea surface. By sight alone the men on watch would have been able to tell nothing of what lay around them.

Even though she is far at sea the *Elise G.* is in a very busy part of the ocean. There are fishing vessels running at night to and from the places where they hope to find fish. There are tankers and freighters bound to and from the ports of New York and Boston, Europe, the Mideast, and the South.

Big ships, tankers and cargo vessels, are very difficult to maneuver in an emergency. Because of their size, it takes them many minutes to turn and many miles to stop. Traveling at full speed, they may cover the distance of three miles in less than six minutes. To be hit by one of these huge ships at night would mean the entire destruction of the *Elise G.* and her people.

To make the *Elise G.* more visible, the bright lights on the work deck have been left burning, but the greatest measure of safety is to be had from keeping a man always awake in the wheelhouse where he can watch the radar. Set to scan the sea at a distance of 12 miles in all directions, the radar will inform the man on watch of the course and speed of any approaching ship. It will warn him if there is danger.

The radar screen is circular. The position of the *Elise G.* is represented on it as a dot at the screen's center. The radar antenna sweeps clear around the horizon every few seconds and with every sweep transmits a radio beam which, if it strikes any solid object on the sea surface, will reflect much as the sound of a voice will echo off a wall. The returning radio echo is picked up by the radar antenna mounted on top of the wheelhouse and will appear as a bright dot on the screen. The distance that dot lies from the center of the screen is equivalent to the object's distance from the *Elise G.* By very carefully watching the dot for a few moments it is possible to tell in which direction it (and the ship it represents) is moving. If the dot moves exactly toward the center of the screen, there is certain danger of collision unless one or both of the ships take immediate action. Kaare has left orders that should any ship headed directly at the *Elise G.* come within 3 miles, the man on watch should start the main engine and run at full speed out of the approaching vessel's path.

But tonight there have been no close calls. At about 3:00 a large ship passed 8 miles to the east, bound toward Europe. Also, there have been six smaller dots on the screen all night about 4 miles to the west, but they have stayed in one small area and have moved only slightly.

These dots, targets as they are called on the radar, represent

other fishing vessels working on this same portion of the fishing grounds. Careful examination of the motion of each of these dots reveals that some are entirely stationary while others move steadily back and forth across a part of the screen, a mile or two each way. The stationary dots represent fishing vessels on which the crew are asleep. The dots that remain in motion are those of fishermen that continue to set and tow their nets through the short hours of the summer night.

The fact that some crews sleep at night while others continue to fish is the first indication that even for boats fishing in the same place, at the same time, there are many different ways of catching the same fish. According to the experience, the judgment, and the whim of each captain and crew, every boat operates in a different way. But all must reckon with the same great forces: the sea, the movement of fishes, and fatigue.

Although the actual routine of each vessel is different, they all have one thing in common: Each is organized so that while at sea the most possible fish will be caught in the least possible time. Most fishermen prefer to spend as little time as possible at sea. Once a trip begins, all energy is directed toward ending the trip quickly and bountifully. This usually means that captains and crews will sleep as little and work as long as they are able. They will continue this grueling routine for as long as it takes to catch a trip of fish. Because of exhaustion and lack of sleep, each succeeding day of a trip becomes more difficult than the one before it.

Four days ago, at about 4 A.M., the *Elise G.* steamed out to sea from Newport, Rhode Island. She headed roughly east to pass south of the Massachusetts coast, the island of Nantucket, the elbow of Cape Cod, and finally into the North Atlantic Ocean. On the morning of the next day, twenty-eight hours from Newport, she set her nets on Southeast Part. Fishing continuously from 9:45 that morning until midnight, she caught 4,000 pounds of yellowtail flounder. Then, from midnight until 4:30 the next morning, the small ship drifted and the men slept.

The next day, at dawn, Kaare steamed to another point on Southeast Part about 15 miles away, hoping to find better fishing. They set the net at 7 A.M. and finished again at midnight. They caught 5,300 pounds of yellowtail, 1,000 pounds of cod, and 600 pounds of haddock; only slightly better than the day before.

During his predawn watch on the third day, Kaare again moved the *Elise G.* to a new position, this time a distance of 6 miles. Fishing from 6 A.M. to midnight, they landed 10,000 pounds of yellowtail, 800 pounds of cod, and 600 pounds of haddock even though they lost four hours of fishing time in mending a torn net.

The *Elise G.*, like all the rest of the vessels on Georges Bank, is allowed in each trip to take 25,000 pounds of yellowtail, 5,000 pounds of cod, and 5,000 pounds of haddock. The quotas are set by the United States National Marine Fisheries Service in an effort to protect these species from overfishing.

With three days' catch already aboard, the *Elise G.* need only catch 5,700 pounds more of yellowtail, 3,200 pounds of cod, and 3,800 pounds of haddock to fill her quota, at which point she will be required to stop fishing and return to port.

Statistically, the trip may be almost over, but physically, the hardest hours are just beginning.

It will be sunrise before the *Elise G.* begins to fish. Kaare drinks his first mug of coffee of the day and watches daylight steadily rise in the east. In the last hour the fog has thinned out. A wind has begun

to gently blow from the distant mainland out to sea. This mainland air is dry and cool. It evaporates the fog and promises a perfect day.

Three stars still pierce the steadily increasing light. To the west, where the radar has reported them all through the night, lie the six other fishing vessels. Red, green, and white lights shine brightly from them; but it is still too dark to make out the shapes of their hulls or details of their rigging. They are only gray objects between a gray sea and an equally gray sky. From the *Elise G.* their lights look like a village in the midst of a vast prairie.

Fifty yards from the side of the ship the fins of two sharks wallow through the water surface. Yesterday there were dozens of sharks, all of them moving to the south southwest; singly and in twos and threes. Kaare watches these two carefully, questioning what directs them. The temperature of the sea? The state of the moon? The length of days in this season of change? Do they choose their own path or do they follow other species, out of sight, swimming depths below?

During the trip before this one, the *Elise G.* fished in this same area for three days. At that time there were no other vessels. She returned this trip to find many other fishermen, mostly from New Bedford, Massachusetts. These men all know one another and the news of a good spot to fish travels fast among friends. Because Kaare fishes most of the year out of Point Pleasant, New Jersey, he knows few of the New Bedford men and in any case he doesn't usually care to fish near other vessels. He keeps mostly to himself.

There are those fishermen who find out about a place where there are fish from other fishermen. They go there and they collect those fish just as a farmer collects grain. They are harvesters.

Kaare is a hunter.

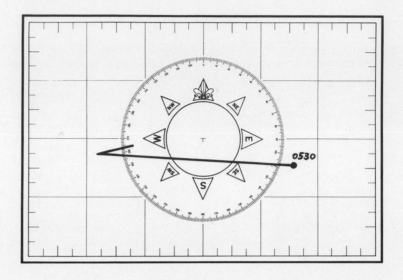

0530

The sky is now bright. The shapes of the vessels grouped to the west are distinct. The stars are gone and it is time for the fishing to begin. Kaare reaches to the instrument console next to the steering wheel and presses firmly on a small, chromed button there. A tremor runs through the vessel, then a deep roar as the main engine turns over once, starts. Kaare watches the engine gauges on the console, First oil pressure, then transmission oil pressure needles rise to their proper positions on the dials. The engine temperature gauge remains motionless for the time being but soon will also begin to rise.

Kaare must allow the engine time to warm up. He leaves the wheelhouse and descends the steep ladder into the living quarters located on the main deck. He steps into the cabin he shares with Tom, the mate. Tom, awakened by the sound of the main engine starting, is lying on his back with his eyes open; Kaare nods to him and goes out. Tom rises and dresses quickly.

Kaare then steps into the other cabin where the rest of the crew sleeps. He sees that Rob is already awake, nods to him, then taps John on the foot and goes out.

Kaare returns to the wheelhouse. Once there he scans the horizon to be sure that there are no other vessels approaching, then swings the wheel toward the group of boats to the west. He levers the engine into forward gear and opens the throttle. The *Elise G.* leaps ahead and turns sharply. Once she is on course, Kaare opens the throttle still farther. In a few moments she is steaming full speed toward the day's work. She runs at 10 knots—which is the same as 11.5 land miles per hour. For a fishing vessel of her size the *Elise G.* is both fast and powerful.

When he is a mile from the rest of the fleet, Kaare reduces speed and swings the ship east. He then steps to the rear of the wheelhouse to look out onto the work deck. He sees that the crew are ready to set the net.

Cut through the stern of the *Elise G.* is a ramp about 6 feet wide. The ramp slopes from the level of the work deck down to the water. Mounted directly over the ramp is a power-driven spool, called a drum, on which the net is wound when taken aboard. By means of a hydraulic motor, the drum may be run in either direction to wind or unwind the net. Because the net is handled through the stern, the *Elise G.* is what is known as a stern trawler. Other New England boats handle their nets over the side; they are called side trawlers.

The net that is used is called an otter trawl. It is a type of net used by fishermen all over the world, designed to be towed across the sea bottom. It is used to catch what are generally known as "ground fish," those fish that spend at least part of their time on or near the sea bottom. New England fishing vessels, whether side or stern trawlers, that are engaged in the ground fishery are generally known as draggers. Those who work on them are called draggermen.

When Kaare has slowed the engine and is headed in the right direction, he signals for the net to be put over. John starts the drum to slowly unwind the net.

The narrow end of the net, where the fish collect, is called the cod end. It comes off the drum first and drops into the ramp. Rob takes a long pole and pushes the cod end down the ramp into the sea. The rest of the net slides into the water as it winds off the drum. Soon it is stretched out astern of the *Elise G*.

The cod end is fitted with a drawstring so that the fish that collect there can be poured out the end of the net when it is brought back aboard. As it becomes weighted with fish, the cod end rubs hard on the sea bottom. It is protected from wearing out by an extra layer of netting, a mat of raveled rope ends, or even strips of ox hide. These devices are called chafing gear and give the cod end a ragged or bearded appearance.

The rest of the net is shaped like a long, flattened funnel. It is about 73 feet long and 100 feet wide at the mouth. The length of each mesh is required by law to be at least 5 inches when fully stretched. This allows small fish to swim through and only marketable fish to be kept. The upper lip of the mouth is reinforced with a steel cable, called a headrope. To the headrope are attached about twenty hollow plastic floats, red, and each about the size of a soccer ball. On the lower lip of the net mouth is another cable, called the sweep. To the sweep is attached a length of heavy steel chain. The weight of the chain holds the lower lip in contact with the sea bottom while the floats along the headrope lift the upper lip of the net as far as possible off the bottom.

The headrope and sweep meet at the corners of the net mouth, where they join at heavy steel rings. To these rings are hooked the tow wires, lines that attach the net to the ship. Hooked into the rings at present are short lengths of rope that secure the net to the drum. But as soon as the net is completely unwound, John and Rob hook the ends of the main tow wires into the rings. Tom winches the wires tight to take the weight of the net off the drum, then John and Rob remove the lines to the drum. The net is now ready to be let out on the tow wires.

John and Rob are at their stations at either side of the stern ramp while Tom stands at the forward end of the work deck. Facing aft, he handles the levers that control the towing winches, located one to his right and one to his left. Tom pushes the two levers forward. The winches start to pay out wire.

From his position Tom can see everything that Rob and John are doing. So long as the work deck is clear he can see the whole length of the tow wires as they come off the winches, stretch aft and slightly upward to the very corners of the stern where they pass through two massive pulleys, called towing blocks, then angle sharply downward to the sea.

The towing blocks are hung at the peak of a massive steel framework that spans the entire width of the stern. This framework is called the gallows. It is built not only to bear the strain of the tow wires, but also to hold the trawl doors.

The trawl doors are rectangular plates constructed of oak and

steel. They attach to the tow wires at a point about 150 feet in front of the net. Each door is attached to its wire so that it will stand on its side on the sea bottom facing at a sharp angle out from the direction it is being dragged. The combination of its skatelike bite on the sea bottom and the pressure of water against its forward-moving surface forces each door to ride far to either side of the ship's track. This forced separation of the tow wires stretches the mouth of the net into a wide grin.

Each time the net is set, the doors must be attached to the tow wires, then cast loose from the gallows where they hang while not in use. When the net is hauled back, the operation is reversed.

Handling the doors at the gallows is the most dangerous operation in the fisherman's routine. Each of the doors on the *Elise G.* weighs about a ton. Whenever they are put over the side or brought back aboard, John and Rob must reach in among the wires, chain, and ponderous steel to handle the chain and hook that secure each door to the gallows. If the sea is rough, the doors will swing and hammer against the gallows like thunder. If the tow wire should slip or the hook fail to hold, the doors crash back into the sea. The strength of a man's arm means nothing to these forces. An accident will certainly be crippling.

Setting and retrieving the net requires very careful teamwork and absolute concentration. The men repeat the same tasks dozens of times a week, hundreds of times a year, and thousands of times in a lifetime at sea. One mistake over that time will mean a man badly hurt, perhaps dead. Every step of setting and recovering the net is done in exactly the same way each time. Each man performs the same operations down to the last detail. By establishing this sort of routine it is more likely that there will be no confusion, no steps left

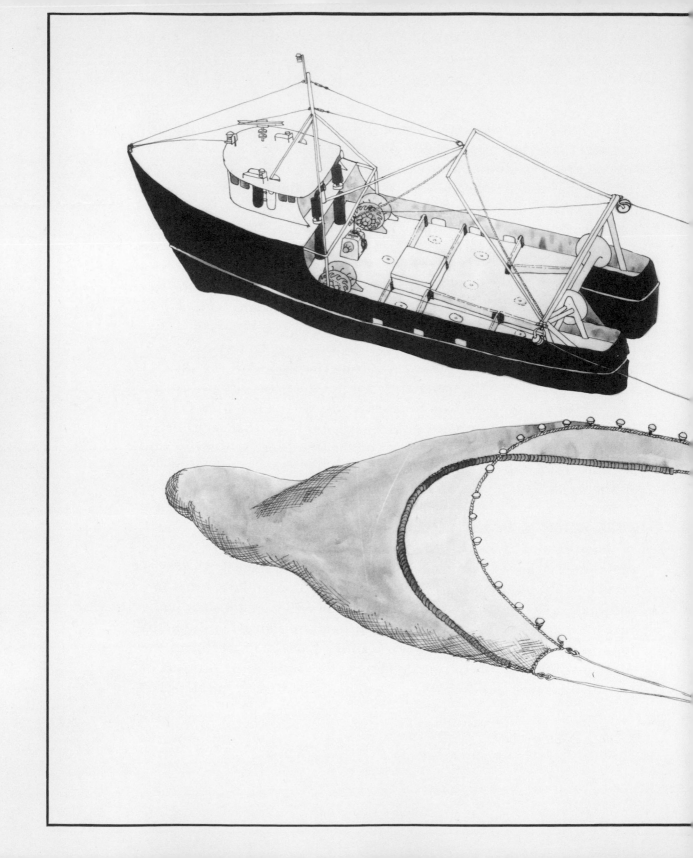

ELISE G.

LENGTH   88′
HORSEPOWER   675
CAPACITY   90,000 LBS. Iced Fish

OTTER TRAWL

LENGTH OF NET   75′
WIDTH OF NET   100′
LENGTH OF WIRE
    NET TO DOOR   425′
    DOOR TO SHIP   750′
    IN 250′ SEA DEPTH

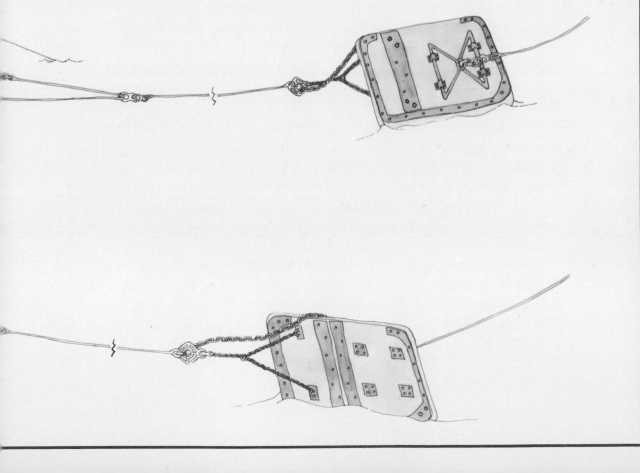

out, and less chance of accident. Tom has become absolutely familiar with the working and the feel of the towing winches. He can make them operate precisely and should be able to tell instantly if there is anything wrong with them, soon enough to warn Rob or John out of the way.

Tom has a great dislike for disorganized work. A few years before, while moving the doors across the deck of another fishing vessel, someone pushed the wrong way. The doors fell together with the end of Tom's finger between them, and the end of his finger was gone. Tom says he was lucky, because if any more of his fingers, his hand, or his arm had been caught, they would have been gone, too. Tom and the rest on board know that the only way to avoid accidents is through constant attention to what is going on around the deck and making sure that each is doing his particular task as carefully as possible. They know that as each day wears on they will become more tired, less able to pay attention, and less able to summon the strength to protect themselves should accidents occur. They know that with each day of the trip they grow more tired and enter into ever greater danger.

When the doors are free of the gallows and touching the water, the net is ready to be actually set. With the ship steaming straight ahead, Kaare increases speed to almost full throttle. The vessel surges ahead and Kaare signals for Tom to begin streaming the wires.

Tom levers the winches to unwind at full speed. The objective is to get the net onto the bottom quickly and to keep as much strain as possible on it while it is sinking. If the net were merely lowered to the bottom from a motionless ship, the doors would tangle with one another and the net would twist, perhaps turn inside out. This is what is known as a "foul set."

"My God," says Rob, "you set foul just once and you spend the

rest of the day and maybe half the night trying to get it straightened out again. You get knots in the wire the size of a pumpkin. Everything's all torn up."

The tow wire continues to stream out. At intervals of 150 feet there are bits of rope fiber twisted between the steel strands of the wire. These tell Tom how much wire he has let out and, according to the depth of water they are fishing in at the time, and how much wire he has still to let out.

To allow the net to sweep the bottom properly it is necessary to stream about three times as much wire as the water is deep. The water where the *Elise G.* is fishing this morning is 250 feet deep; this means 750 feet of wire. Tom carefully counts the markings as they flash off the winch; when the proper length is near Tom slows the winch and, with a hammer that he keeps by the winch controls, strikes a loud ringing blow against the steel. This is the signal for Kaare to slow the engine which, abruptly, he does. With the strain on the wires reduced by the slowing of the engine, Tom brings the winches to a stop so that the wires are at exactly the same length. John and Rob then set the locks on each winch and disengage their motors so that no more wire may be pulled from them. Tom then strikes twice more with the hammer to tell Kaare that all is ready for towing.

Kaare brings the engine back to about two-thirds speed. At this engine setting the ship would normally run at 7 to 8 knots, but she now makes barely 3. The drag of the net through the water and the strain of the doors digging and shearing at the bottom 750 feet astern and 250 feet below, give the ship all she can do to move ahead. Another day's work is begun.

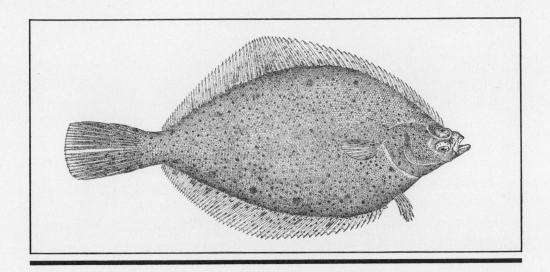

Yellowtail

LIMANDA FERRUGINEA

Before 1930, no one cared much about the yellowtail flounder. The species is not even included in some major nineteenth-century works on American fishes.

Up to that time the only flatfish able to attract a following was the halibut. And halibut were astonishing brutes—voracious, fast swimmers, growing to 9 feet and 700 pounds. They could be caught on baited hooks anywhere on the northeast coast of America from close inshore to the very edge of the continental shelf.

But by the time the otter trawl had come into common use, the number of halibut had been so depleted that they were only an incidental catch. Cod and haddock were the first choice of the draggermen. But with each tow of the net came many, many, smaller flatfish. There were lemon sole and winter flounder, gray sole and sand dab, northern fluke and yellowtail flounder (also known as rusty dab). Inshore fisherman had been collecting flatfish for market wherever they were able. People liked the flavor and the clear, white, boneless meat. By the 1930s a major fishery had developed for yellowtail in the offshore waters of southern New England, to the south and west of Georges Bank.

In comparison to its giant cousin, the halibut, a yellowtail flounder is a small fish. It grows to a length of only 2 feet and a weight of maybe 4 pounds. But while the halibut were found only in small roving packs, whole populations of yellowtail could be found on yellow sand bottoms at depths of from 30 to 300 feet.

Yellowtail demand clear, cold ocean water. They spawn between May and July, when the water temperature at the sea bottom ranges between −0.5° and 8° Centigrade. Although the fish spend their lives close to the bottom, their eggs float at the surface for five to ten days before hatching out into free-swimming larvae that look not much different from the hatchlings of any other fish species.

But as the young flounder grows, changes take place. The eye on the left side of the head begins to migrate up and across the top of the skull until both eyes lie together on the fish's right side. The fish at the same time flattens as if it were a lump of putty lying on its side in the sun. Soon the fish has remodeled itself to lie on the sea bottom flat on its side, both eyes pointing upward. The bottom, or eyeless, side of the fish becomes pure white except for two smudges of bright yellow beside its tail. The top side, the sighted side, assumes a pebbled brown and rust-red coloration to blend with the sea bottom where it will spend its life once the transformation is complete.

The Latin word for "right" is *dexter*. "Left" is *sinister*. Yellowtail normally develop with their right sides uppermost. For this reason they are called "dextral." Other species of flatfish, such as fluke, develop with their eyes and coloration on the left side and are therefore called "sinistral." Sometimes the process for an individual becomes confused and a normally dextral yellowtail will develop as a wholly sinistral fish. Far more frequently, a confusion of body coloration is seen. The back of an otherwise dextral fish may be blotched with sinistral white while the underside will bear patches of brown and rust-red.

In 1949, records show that yellowtail began to appear in numbers on Georges Bank and at the same time began to grow scarce on the southern New England grounds. Soon after that a few began to be taken off the coast of Nova Scotia, far to the north and east. Clearly some drastic change was occurring. Fisheries biologists tried to find out why.

It had long been known that yellowtail on the southern New England ground tended to migrate toward the east in the summer and back again in winter. This migration usually extended only about 50 miles, but by tagging individual fish, releasing them, and then seeing where they were later caught by fishermen, the biologists found that some fish made seasonal trips of up to 170 miles. The fact that this migration was seasonal and toward the colder waters to the east suggested to the scientists that high water temperature might be what made the fish move.

Another researcher began to dig through weather bureau re-

cords for the New England coast. He found that between 1945 and 1955 there had been a considerable warming of average air temperature. Certainly, if the air got warmer the seawater ought to have warmed a few degrees, too.

But could this small increase produce such a drastic shift in the yellowtail population? The observation that even before the temperature increase the fish were forced to move east in the summer suggests that it could. It now seems quite certain that the waters of the southern New England ground before 1949 were only barely cold enough to provide a usable habitat for the yellowtail. A very slight increase forced the entire population to abandon an area of thousands of square miles. Presumably, if the temperature dropped slightly for a period of years, the yellowtail would be able to move back. If the temperature rose even higher, they might be forced to leave Georges Bank as well and move even farther north.

Few realize the true delicacy and ambiguity with which one thing rests upon another.

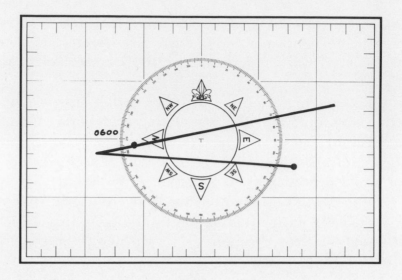

0600

The sun is now five minutes above the horizon. The hulls of the vessels nearby are bright and the air is cool. Tom, Rob, and John step back into the living quarters. It is time to get breakfast and wait the hour and a half until the net must be hauled back.

The living quarters are by a fisherman's standard very comfortable. They are located in the high bow of the vessel under the wheelhouse. They are completely air-conditioned so that the air there is even-temperatured and always dry. The largest space is the galley, where the cooking and eating are done. The galley is equipped just like a home kitchen with a large refrigerator, cupboards, a sink with hot and cold water, and an electric range. The top of the stove is fitted with a puzzle of adjustable bars to keep the pots and pans from sliding about as the ship rolls.

The galley table, where the crew eats, is surfaced with carpeting so that the dishes will not slide. Built in the center of the table is a rack to hold salt and pepper shakers and sugar bowl. Also on the table are numerous odds and ends, a pack of cards, several packages of chewing tobacco, a portable tape recorder and a dozen tapes, a few odd nuts and bolts, and a battered hat. Around three sides of the table is built a deep-cushioned settee.

Eating at this table in rough weather is an acrobatic act. One hand must hold a bowl (plates are not used because the food would slide off them). The other hand, between dashes to the mouth with a fork, must be ready to catch any objects that start to slide off the table. But meals go on pretty much as usual even in rough weather. The fishermen only rarely become so seasick that they feel unable to eat. And even though they often feel queasy in bad weather, they testify that the quickest way to feel better is to eat a good meal.

Adjoining the galley is the bathroom, or "head" as it is called aboard ship, and two bunk rooms. The head is equipped with a toilet, a washbowl, and a shower with hot and cold running water. There is a medicine chest choked with razors, Band-Aids, toothpaste, Alka-Seltzer, and aspirin.

The two bunk rooms have a total of six double-decker bunks. Kaare and Tom share one room, John and Rob the other. The boat was built with many extra bunks because she might at some time be used in a different type of fishing requiring a larger crew. Because the ship rolls constantly at sea, the crew prefer the lower bunks near the centerline of the ship. The motion is less there. All of the bunks are faced fore and aft. (A bunk faced to the side is almost impossible to sleep in at sea because the person lying in it constantly slides back and forth, alternately hitting his head and then his feet against the ends.) The bunks are fitted with high sides so that one is not so likely to be tossed out in rough weather.

There are no portholes or windows of any sort in the living quarters. When the electric lights are turned out in the bunk rooms, it is absolutely dark. Upon waking up, it is impossible to tell if it is night or day. One must walk into the flourescent-lit galley, then look out the door onto the work deck, to tell what is happening outside. There is a unique sense of security and comfort in being so cut off from the sea outside. After being on the work deck for twenty hours, the crew have usually had quite enough of the sea air. They are glad to get as far from it as possible.

The *Elise G.* is a new and very modern vessel. Kaare insisted that she be made as comfortable as possible. Until very recently, no owner of a New England fishing vessel would think of installing air-conditioning or a shower, and any sort of toilet other than an ordinary bucket was uncommon.

In older vessels the crew sleep in the forecastle ("foc'sle"), the cramped, poorly ventilated space below deck in the very point of the bow. All cooking is done on an oil-fired stove in the same tiny space. When the stove is off, the forecastle is often cold and damp, smelling of mildew, sweat, and fish. When the stove is on, it is terribly warm and smells of rough food and hot oil. Overall, it is like living in a defective incinerator.

Rob, like everyone else on the *Elise G.*, has fished on plenty of other boats, and he can tell a good bunk from a bad one. "Man, you're on one of those boats and you look around and it's like an oven and everything is covered with a half an inch of grease and nothing's been cleaned or even swept for about twenty years; and you've got six or eight guys all living down there. You look around sometimes and ask yourself, 'What am I doing here?' "

Kaare will tell you that one of the main reasons he chose a vessel like the *Elise G.* is comfort. "You know that if you eat better and you sleep better, then you got to catch fish better, too. Those other boats, you know, I went on them for over forty years and I caught a lot of fish on them, too, but I never caught as many fish as I can with this one. Because you got to have sleep and food or else you get too tired, you don't catch so many fish, and maybe you hurt yourself."

Just as every man has certain duties in setting and recovering the net, so there are duties involved with living aboard the *Elise G.* Kaare dictates when the boat fishes, when the crew sleeps, and who must stand watch. Rob cooks all the meals. Rob, as cook, makes out the shopping list for each trip; John shops. John also sweeps and mops all the decks in the living quarters and the wheelhouse at the end of each trip, and Rob scrubs the stove and sink. Kaare and Tom share the work of cleaning the engine room and attending to the regular maintenance of the generators, pumps, and winches.

Some fishermen eat well while others eat poorly. There are good sea cooks and there are bad sea cooks. Vessels with a reputation for good food are called "fat boats." Others are called "hungry boats." It is said that seagulls don't even bother to fly for the scraps tossed from a hungry boat.

The *Elise G.* is a fat boat. Rob begins breakfast by scrambling a dozen and a half eggs, frying half a pound of bacon, and toasting half a dozen English muffins. He makes a pot of oatmeal and refills the coffee pot. He makes fruit juice and quarters a melon. In a short time, all of this food is set out on the galley table. John and Tom sit down to eat. Rob climbs up to the wheelhouse to run the ship while Kaare comes down to eat.

"Good," says Tom.

John says, "We're all gonna be so full we'll need to take a nap."

Kaare looks up questioningly. "What, you mean you didn't get enough sleep already?"

John is the youngest and the least experienced on board. He finds it very difficult to joke with Kaare. Rob does not often attempt it either. Tom is the only one who dares, and even he does it with considerable caution.

The authority of the captain of a fishing vessel is a very subtle thing. In the navy or the merchant marine all levels of rank are set out according to long-standing tradition and law. Orders travel down through these levels as rigid commands which are intended as much to insure that the crew and the ship will be protected from any whims or oddities of the captain's personality as they are to enforce the will of the commanding officer. The behavior that the captain of a warship must observe toward his men is just as tightly regulated as the obedience that the crew must give the captain.

On a vessel such as the *Elise G.* the captain's word is no less the law, but very often the actual words become difficult for the crew to interpret. Many of Kaare's most forceful and important wishes are communicated by a nod, a growl, or a hard-eyed silence.

While an enlisted man in the navy has no alternative but to obey the commands directed at him until his enlistment ends, a crewman on a fishing vessel has a free choice of whether or not to sail on any trip. A captain who becomes known for making foolish or unreason-

able demands on his crew will be unable to man his vessel. However, as long as he does nothing to insult the pride of his men and so long as his authority is directed solely toward the business of catching fish and staying alive, the crew will willingly labor under the tightest sort of rule.

While the majority of the workers ashore would probably like some day to own their own business or for once be the boss, this tends not to be the case with fishermen. Most fishermen, even though their experience at sea may span several decades, have no desire to undertake the responsibilities of captain or owner of a vessel. A deckhand makes good money, only slightly less than the captain. Those who command do so because, first, they want to, and second, because they are permitted authority by the men under them.

Anyone who chooses a life at sea, early comes to understand that there are times, very desperate times, when there can be only one voice. In emergency the very most important thing is that whatever action is taken, it be unanimous. It is quite likely that out of any tight situation there will be more than one possible escape. But even a moment of indecision or, worse yet, dispute as to which means of escape should be tried, will spell disaster for everyone.

At sea there can be only one captain. If a man finds that he has begun to challenge the captain's authority, it is time for him to get his own ship. If a crew begins to doubt the captain's ability, they had best get off as soon as they can.

When Kaare is fishing, he talks very little more than what is required for good manners and the running of the ship. If a conversation seems to him to be getting off the point, he will as likely as not simply turn on his heel and go back to whatever he was doing.

Kaare works with absolute concentration. He drinks as many as twenty-five cups of coffee a day, sleeps little, and when he does lie down he remains fully dressed. But as soon as the last tow of a trip is completed, Kaare is transformed. The diamond-hard tension dissolves. He holds forth in the wheelhouse with stories of fishing and men during the long run back to port.

This morning as always, Kaare is first to finish his meal. He goes to the stove, refills his mug, and climbs back to the wheelhouse to relieve Rob, who comes down to eat. The rest of the crew have left a good share of food for Rob even though all of it is by now stone cold. But Rob is used to cold food. He, after all, is the cook, and on this boat, the cook eats last.

As soon as everyone is finished eating, Tom begins to clear the table and John starts to wash the dishes. The time when the net must be hauled back is fast approaching, and the galley must be cleared up by then because there will be no time to do it later.

Kaare tows the net east. Behind him the tow wires angle sharply to the left as they run from the stern of the ship to the sea surface. At first you would think that the boat must be turning or going off course, but a check of the compass by which Kaare is steering shows that there has been no change in the ship's heading. Still the net is obviously tracking far to the left of the ship. Kaare frequently glances out through the windows in the rear of the wheelhouse to be sure that the angle of the wire has not increased.

That the ship and the net are towing crookedly is the fault of a strong tidal current. Just as the tide flows in and out each twelve hours along the coast, so the waters of Georges Bank are also in motion. But while the currents close to shore normally flow only in one direction, then slow down, stop, and start to flow six hours in the reverse direction, the tidal current offshore constantly flows at about the same speed and gradually shifts in a clockwise direction until, in twelve hours, it completes a circle of 360 degrees.

At 5 A.M. this morning the current here was flowing east. By 8 A.M. it will be flowing due south and by 11, west. By 2 in the afternoon it will flow north, and so on—all at a speed of about 1½ knots.

Because the *Elise G.* is towing the net at a speed of only 2 to 3 knots and because the current is at this moment flowing across her path, for every 3 feet she moves forward she slides 1½ feet to the side. The net, on the other hand, is not so much affected by the current because the doors are keeping a firm grip on the bottom. With the ship and the net trying to move in separate directions there can be problems.

If the angle becomes too great, the lefthand door is likely to lose its grip on the bottom and the mouth of the net will fall out of shape. Kaare cannot see the net as it is working on the bottom, and he has no way of knowing for sure whether the net is fishing properly. He can only allow for the tide by watching the angle of the wires coming in over the stern. If the net goes out of shape, he will find out what has happened only when they haul the net back and find it empty.

It requires his long experience to imagine what the net must be doing on the bottom. Kaare knows every part of the net by heart—its ropes, wires, floats, twine, and chain. He built the net himself last winter. He knows how the doors are supposed to act on the bottom both from his own experience and from dozens of technical and scientific reports he has read in his lifetime. But still he has never actually watched his net, or any net, as it fishes.

Two hundred-fifty feet below the surface it is very dark. Only 1 percent of the sun's light penetrates to that depth. What little light does get there is a deep, strong blue.

Perhaps a fish at the bottom can hear the sound of the *Elise G.*'s engine and propellor laboring overhead, but even if that sound can be heard it is long past by the time the doors approach.

The doors are about 250 feet apart. They grate and rumble as they churn through the bottom. The wires that run back from the doors to the mouth of the net whisper as they slide through the sand. Any fish that happens to find itself between the doors as they pass is alerted. Danger lies to either side, and the fish will choose to swim to a point an equal distance from each door. This will bring it into the path of the net. The doors pass, leaving a trail of stirred up sand and mud like a smoke screen behind them. The wires chatter and hiss within this black cloud.

The fish hesitates. Its hearing is excellent. Its eyesight, developed to work in this blue darkness, is perhaps one hundred times more sensitive than man's. Through special organs in its body, the fish can sense the movement of distant objects by subtle changes in the water pressure around it. All of these senses are now in alarm.

The wires on either side draw closer. The sound level increases; the commotion of the bottom from the wires on either side is now plainly visible. And from behind comes a new sense of menace.

There is vast movement. Movement and sound. The chain on the lower lip of the net scrapes and rattles across the stones. The great shroud of twine in the bag of the net sighs against the flow of the sea through it. The upper lip, perhaps 10 feet off the bottom, sweeps over the fish like a cloud across the moon. The wings of the net block escape to the sides and for moments the fish swims frantically in the mouth of the net, until in panic and exhaustion it turns wildly down the twine throat. It strikes the twine, recoils, and strikes the twine on the other side of the ever narrowing belly. Finally it lodges in the

cod end, held tight and helpless against the twine and crushed by the weight of more and more fish that enter the net.

Everything that rests on or swims near the sea bottom winds up in the mouth of the net—valuable fish and worthless fish; lobsters and crab; sticks, shells, and small boulders; rusty cans and throw-away boots. Things that are small enough pass right through the mesh of the net and escape.

As the net fills, it becomes increasingly difficult for the ship to pull forward. The increased drag and the slower forward speed pull the doors closer together. The mouth of the net eventually begins to go out of shape and the net fishes poorly. Also, the more material that collects in the cod end, the greater the crushing pressure on the valuable fish there. A fish that is crushed in the cod end or "wrung" in the twine of the net will spoil sooner in shipment to market. A fishing vessel that tries to sell damaged or spoiled fish will have to settle for a lower price or perhaps no price at all.

It is important to haul the net back before it becomes too full. The length of time the net can be towed depends on the quantity of material that is being caught. Sometimes a vessel will tow as little as half an hour to get a full bag. At other times three hours will not be enough. Yesterday the *Elise G.* was making hour-and-a-half tows, which seemed to be about right.

Kaare leaves the wheelhouse and climbs down the ladder, with one hand gripping the rungs and the other gripping his coffee mug. He enters the galley, goes to the stove, and refills. The crew are sitting around the galley table, each engaged in some idle pursuit. Kaare nods to them and goes out. It is time to haul back.

Just inside the door that leads out to the work deck hang the crew's waterproof clothing. Because these garments were once made of heavily oiled cloth they are still called oilskins, even though they are now made of a more durable plastic/rubber material. They consist of a pair of high rubber boots, a pair of coveralls held up with suspenders, and a jacket that comes down to the hips and snaps in the front. To cover the head there may be either a hood fitted to the jacket or a broad-brimmed rubber hat called a sou'wester. The sou'wester appears old-fashioned in comparison to the hood but, as Rob points out, rain and spray have changed not at all over the years, and a hat that worked well one hundred years ago ought to work well today.

Because the day is fine, the men need only to protect themselves from water running off the net and from the fish they will be handling. For this they will need only boots and coveralls. These they put on, balancing expertly on first one foot and then the other as the deck rolls beneath them.

Once dressed, Tom, Rob, and John pull on thick rubber gloves to protect their hands. Not all fishermen wear these gloves. Many use thin cotton gloves because they are less bulky and permit a better grip on small objects. A few wear no gloves at all, whether it be winter or summer.

Constantly soaked in salt water, bruised, scraped, jabbed, cut, and alternately frozen and burned a fisherman's hands are the hardest worked and most highly developed of any profession where strength, dexterity, and resistance to injury are combined.

Most fishermen are proud of their hands for the grueling work they are able to perform with them. Tom says, "You hear musicians and surgeons all the time talking about how they have to be careful and how they've *got to take care of their hands*." He holds his hands open palms up before him "Hell, lookit, I don't have to take care of these hands at all. They take care or me!"

The Chart

Fishermen have found their way to Georges Bank for more than a century. Coming from Gloucester and Boston the old schooners sailed south-southeast 150 miles, until with weighted lines they were able to measure the depth to the bottom on the northeast edge of Georges Bank. From there they steered more to the south and ran the 20 miles farther to the Winter Fishing Ground. They took more soundings as they went to be sure they were safely on course.

They compared the measurements they took with figures printed on their charts. Even an average captain could tell you the

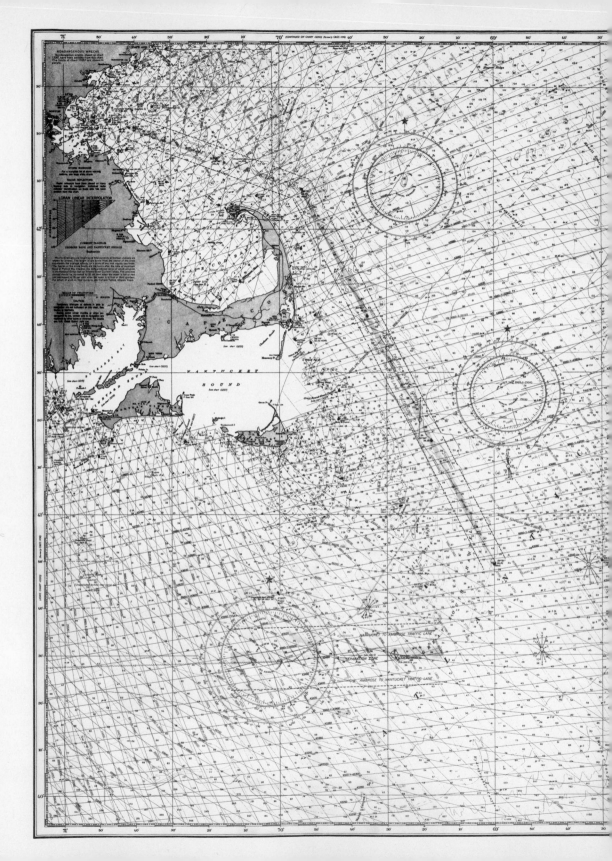

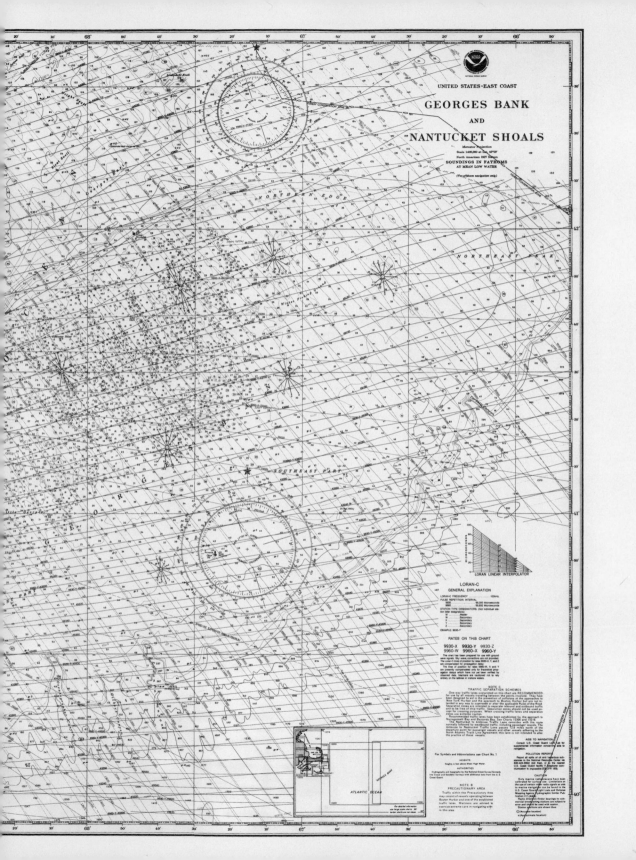

UNITED STATES-EAST COAST

GEORGES BANK
AND
NANTUCKET SHOALS

Mercator Projection
Scale 1:400,000 at Lat. 40°00'
North American 1927 Datum

SOUNDINGS IN FATHOMS
AT MEAN LOW WATER

(For offshore navigation only)

LORAN LINEAR INTERPOLATOR

LORAN-C
GENERAL EXPLANATION

LORAN-C FREQUENCY 100kHz
PULSE REPETITION INTERVAL:
   9930 .... 99,300 Microseconds
   9960 .... 99,600 Microseconds

RATE TYPE DESIGNATORS (Not individual station letter designations)
   M .... Master
   W .... Secondary
   X .... Secondary
   Y .... Secondary
   Z .... Secondary

EXAMPLE: 9930-Y

RATES ON THIS CHART

9930-X   9930-Y   9930-Z
9960-W   9960-X   9960-Y

NOTE C
TRAFFIC SEPARATION SCHEMES

For Symbols and Abbreviations see Chart No. 1

NOTE B
PRECAUTIONARY AREA

ATLANTIC OCEAN

texture, the color, the smell, and the taste of the bottom on any part of Georges Bank.

As they returned to port in snow or fog the schooner captains might again take soundings to discover when they were near land. It is said that one early captain tapped his way over the sea bottom like a blind genius with a cane.

Kaare uses chart number 13200 (Georges Bank and Nantucket Shoals) printed by the United States National Ocean Survey. On it are many of the same sounding figures that the schooner captains used. But there is much more. The chart is a detailed description of everything known about the terrain of the Bank. The locations of shoals, wrecks, and lone rocks are shown. Kaare can measure the direction and distance to any point within half a compass degree and a fifth of a mile. With an electronic depth recorder he can take continuous soundings precise enough to show not only the bottom but also schools of fish swimming in mid-depth. And, like all of the Georges Bank fleet, the *Elise G.* is equipped with LORAN, an acronym which stands for *Lo*ng *Ra*nge *N*avigation.

This system is tremendously accurate and requires the fisherman only to purchase a special radio receiver capable of picking up LORAN signals. The rest of the system is on land and provided by the government.

Two transmitter towers are placed far apart on the coast. At an exactly measured time interval, first one tower then the other broadcasts a signal. These two radio signals travel out to sea at the speed of light. If you are located closer to one tower than the other your LORAN receiver will pick up the closer signal slightly sooner than expected. The receiver is built to measure the time difference between signals to an accuracy of one *millionth* of a second.

The LORAN receiver on the *Elise G.* automatically reports on two pairs of signals at once. It flashes two sets of numbers which correspond to numbered lines printed on the chart: for instance, 1H3-2397 and 3H4-6278. Kaare compares the figures to find his position on the chart.

Because LORAN is now so widely used, the fishermen no longer speak of parts of the Bank by their old names such as Cultivator Shoal or Little Georges. They identify places by their LORAN positions. To make a trip to Southeast Part is now to "go out on the 6200 line."

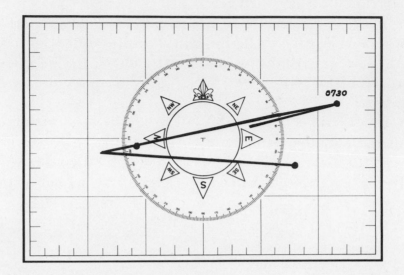

0730

The crew are on the work deck to haul back the net. Tom is at the controls of the towing winches. John and Rob are at their stations between the legs of the gallows. Kaare slows the engine, Tom heaves back on the winches' lever controls, and the wires start to reel in.

The wires are under great strain. As they wind onto the winches they shift; one strand slips across another with the sound of a pistol shot. The markings on the length of the wires one after the other reappear from the sea. Rob and John peer over the side awaiting the first sight of the doors as they rise from the depths.

Each man holds in his hand the hook with which he will secure the door to the gallows. As soon as the door comes in sight he raps the hook against the gallows steel to warn Tom.

The doors rear out of the sea shedding great falls of water. They rush to the peak of the gallows. Tom freezes them there with a stroke of the lever so each is pulled tight into the towing block. There he holds them while John and Rob reach in to attach the hooks. Then he relaxes the brakes. As soon as the doors are hung solidly from the chains and all of the strain is off the tow wires, John and Rob again reach in to unclip and the hauling back of the wires resumes.

Yards astern the bright red floats on the upper lip of the net break the surface. When the corners of the net are in reach, the two men lean far out over the stern to clip on the lines of the net drum. Tom then slacks away the tow wires until the drum has all of the strain, and John starts the drum winding in. As soon as the ends of the tow wires are in reach they are unclipped and the net is ready to be wound onto the drum.

Tom leaves the winch controls and walks back to operate the drum. Kaare comes down from the wheelhouse to a set of steering and engine controls at the front of the work deck. From here he can see better what is going on and maneuver the ship more accurately.

The head and ground ropes wind onto the drum and then the yards and yards of netting that make up the body of the net. Where the belly of the net has been brushing against the sea bottom it is festooned with dark brown seaweed. Here and there a small fish or a squid has become tangled in the mesh and is wound onto the drum, buried and crushed under succeeding layers of twine.

The chain, the rings, and all of the metal on the lower lip of the net as well as the steel runners on the bottoms of the doors are polished bright silver by sliding on the sea bottom. Clear, cold water cascades from the net as it is hauled up the ramp and wound onto the drum.

Sea birds gather. There are tiny petrels, small as swallows, that flutter and pick at the sea surface, dabbling their feet as if walking on the water. They are named after Peter, the saint of Christianity who walked once briefly on the sea until he sank from self-doubt. There are larger birds, shearwaters, with wings like perfect blades that sometimes slice the face of a wave, fast and a fraction of an inch from the water, then wheel and loft into the sky in a banking turn. This far from land there are no ordinary gulls.

The net becomes more and more narrow until finally the cod end appears at the bottom of the ramp. It is packed and bulging with fish, so heavy that the drum is unable to lift it out of the water.

Rob and John come around in front of the drum to stand at the top of the ramp. They pass a line around the net just above the cod end and hook the line to a block and tackle (a system of pulleys) that hangs from the rigging above. With this block and tackle they will pull the cod end aboard.

The *Elise G.* is now stopped dead in the water. Her engine is no longer in gear. Kaare wraps the hauling part of the block and tackle around the capstan of the winch and draws it tight. The capstan is a revolving steel drum that is mounted beside the winch controls. A line wrapped around it and pulled tight will hold to the drum by friction. As the drum revolves, the line is pulled in. To maintain this friction, the man holding the end of the line need only keep a firm pull by taking in the rope hand over hand as it comes off the drum. If he desires to stop the line from coming in, he need only relax his pull slightly and the line will slip on the drum. By letting the line slip he may either hold a heavy object stationary or let it down. By merely pulling the line tight he may hoist again. With some practice it is possible to manipulate extremely heavy objects with great precision.

The cod-end is pulled up the ramp and onto the deck. As soon as he can, John grabs hold of the drawstring that will open the end of the net. It would not do to have the net open too soon and dump the

whole catch of fish right back down the ramp, so as Kaare continues to hoist, Rob wrestles the great weight until it is positioned squarely in the middle of the deck. John then tugs sharply at the drawstring. The end of the net bursts open and the contents rush out. Rob clings to the netting over his head as his feet are swept from under him by the deluge of fish pouring across the deck. He thrashes his legs to regain a footing. The rush of fish sounds like a mud slide all alive with the slapping of thousands of tails, a strange croaking sound, and the clatter of clam shells. The air is filled with a rich, slightly sour smell that is perhaps the odor of a thousand agonies.

In an instant the net hangs empty and limp as a sock. Rob and John shake it once or twice to dislodge any fish tangled in the twine, then Kaare lowers it back to the deck while Tom winds it onto the drum. The purse string is retied as Kaare climbs back to the wheelhouse. He swings the *Elise G.* around in the direction she has come from, gives the signal, and the net goes back into the sea.

The first tow has been a good one and Kaare will tow right back through the same area. The net is back on the bottom within five minutes of the time it was hauled up the ramp. Any fisherman knows that you catch no fish while the net is on deck.

The net has left about 4,000 pounds of fish and debris piled in the middle of the deck. Included in the pile are yellowtail flounder, cod, haddock, several other species of valuable flounder (lemon sole, gray sole, dab, winter flounder, perhaps a small halibut) plus anglerfish, scallops, a few lobster, and a butterfish or two. All of these will be kept and altogether amount to about 1,000 pounds. Everything else is thrown, shoveled, kicked, and hosed back over the side.

Tom, Rob, and John take up short wooden sticks. At a right angle from the end of each protrudes a sharp steel spike. These tools are called picks and are used to stab the valuable fish, lift them, and toss them into clear areas of the deck to either side of the pile of fish. These clear areas are walled off from the rest of the deck by boards about 12 inches high. The cod and haddock are tossed to one pen; all of the flounder are tossed to another. Big anglerfish are heaved into still another corner, while the few lobster, scallops, and butterfish are separated into wire baskets.

The men work quickly. They try to stab each fish in either the head or the tail so as not to damage the flesh of the body. They sort carefully through the pile to be sure they miss nothing. To bring even one valuable fish to the deck only to toss it back over the side is unforgivable.

The picking takes about twenty minutes. Like all the rest of the work done on deck, it is done from a stooping position which for the back of the beginner is extremely painful. But the crew of the *Elise G.* is used to it and work all day bent over so that their finger tips at any time might touch the deck.

Finally there is nothing left in the middle of the deck but trash. There are many small sharks known as dog fish. There are crabs and small rays, called skate. There are sculpin—small bony fishes with long spikes from their heads so sharp they easily pierce the men's boots and stab into their feet. It is these nasty little fish that utter the croaking sound heard when the net is being dumped.

There are bits of junk, scraps of metal and wood, bottles, tin cans, stones, weed and clam shell. Sometimes there are bones, the skull or vertabra of a porpoise. There may be strange fishes. Yesterday the net brought up a large sea turtle. The crew helped it back

over the side, and it swam off across the surface as fast as it was able, its head out of water, looking back.

Sometimes they encounter a great black flabby fish called a torpedo, which weighs as much as 200 pounds and, like an electric eel, may give a violent shock to anyone that touches it. The torpedo shocks the fish around it in the pile. The voltage causes them to leap and twitch wildly. The crew have learned after a few surprises that these monsters may be held by the tail without danger and in this way they are heaved back over the side.

Now and then the net will take a large shark—10 feet or more. Being careful to avoid the shark's jaws, the crew must loop a rope around the shark's tail and hoist it back over the side. They could try to kill it with a blow to the head, but there is really not much point in this. There is little one can do to really hurt a shark. Its brain is so small and reflexes so tough that the hardest blow will at best only stun and at worst cause the shark to go into a fury of thrashing. Usually it is best to leave well enough alone and to slip the beast back into the sea as quickly and as gently as possible.

Rob worked for several years collecting marine specimens for colleges and laboratories. He can identify most of the creatures that come up in the net. As he picks through the pile he will sometimes stop to examine some animal, perhaps pick up a pink sea worm or some species of clam. In this pile of trash he finds an animal about the size and shape of a butternut. Its back is covered with long, soft hair. Its underside is delicately organized, soft ridged and tan. It is a species, he says, of annelid worm. It is named *aphrodite* after the Greek goddess of love and beauty. Rob examines it carefully, then puts it back into the sea.

Winter Fishing

On the East Coast of the United States, from November until May, a constant succession of storms develops off the Middle Atlantic states, especially in the area of Cape Hatteras. From there the storms travel in a northeasterly direction, sweeping along the coast, across Georges Bank, and on to Nova Scotia. The storms are spaced roughly three to four days apart, but there may be less than a day between storms or as many as ten days. Each storm varies in intensity. One may blow only 20 to 30 miles per hour and go almost unnoticed while three days later the next one may gust to 100.

The intensity of a storm may persist for ten to twelve hours. Some storms just die out; others are finished by a last furious succession of squalls so intense that they flatten the seas and tear great clouds of spray from the water's surface. But sooner or later the wind shifts into the west from which direction it may continue to blow for two days or more. The cloud cover begins to break shortly after the wind shifts. The sky is revealed to be an icy blue and the temperature falls dramatically.

The west wind, even after the sky has cleared, is often the most dangerous part of the storm. With very low temperature and a rough sea, great quantities of spray freeze on a vessel. The jacket of ice may grow to a thickness of several feet and the weight becomes immense. The vessel is very likely to roll over under this weight unless the ice can be hammered and chopped away as it forms.

There are many accounts of desperate fishermen battling to reach port against an icy head wind. Every mile they gain toward safety adds still greater weight of deadly ice. The grim game has been won by no more than a few hundred yards and a few frozen pounds. As often, it has been lost by as much.

A coal miner faces the greatest risk of death of any land-based worker in America. But every day a fisherman works he is 2½ times more likely to be killed. And the risk of death is not a random one at all. Each day, a man's survival depends entirely on his attention to his work, his surroundings, and his fellow crew members. The risk is vastly greater in winter than summer, so much so that many vessels do not fish at all in winter or else work close inshore where they can run for shelter with the approach of bad weather. But because there are fewer vessels working in the winter, the market price of fish is usually higher, the temptation to risk a little greater.

A situation in which the risk is thought to be at a tolerable level is called "safe enough." A situation where the risk is judged to be too great is called simply "no good."

Boats that go to Southeast Part in the winter, often go in pairs and try to keep within an hour's steaming time of one another. Most captains figure that whatever trouble might develop they should be able to keep afloat for at least an hour with good pumps. By sticking together when they are far from shore, each crew has a better chance of survival.

During the winter the *Elise G.* fishes out of Point Pleasant, New Jersey. From there she makes trips as far south as Cape Henry in search of porgies.

Porgies are relatively small fish. They rarely grow to more than 2 pounds. During the summer they inhabit shallow coastal bays and rivers where they are caught in numbers by sport fishermen. As winter approaches, the fish move offshore and gather into dense schools. These schools migrate southward as winter progresses; then they turn back northward in time to return to the summer shoreline. Although a school of porgies may remain in one place for several days at a time, it might also move 10 miles in the space of a day. On the way south the schools stay in water of 200-foot depth. On the return trip they often resort to depths of 300 feet, apparently to avoid the winter cold water that pours out of the bays and rivers along the coast and, which because it is more dense than the warmer ocean water, tends to spread out across the bottom.

Unlike yellowtail fishing, where each trip is made up of an endless series of almost identical tows, porgy fishing is more a matter of cunning and stealth. The fishing, when it comes, is usually brief and often spectacular.

Kaare knows that porgies tend to congregate on the down-current side of seabottom ridges. He knows also that the fish will tend to move toward the sun as it crosses the sky during the day. (One explanation for this movement is that the fish can more easily see the small organisms on which they feed when they orient themselves so as to keep their prey silhouetted against the sun's light as it penetrates to the bottom.)

When Kaare arrives at a place where there ought to be porgies, he begins to search the bottom with the depth sounder. Not only will this machine show him how deep the water is and where the ridges and gullies are, it will also record the presence of schools of fish as the ship passes over them. Patiently, Kaare prowls at the surface. By constantly referring to the depth sounder, the chart, the compass,

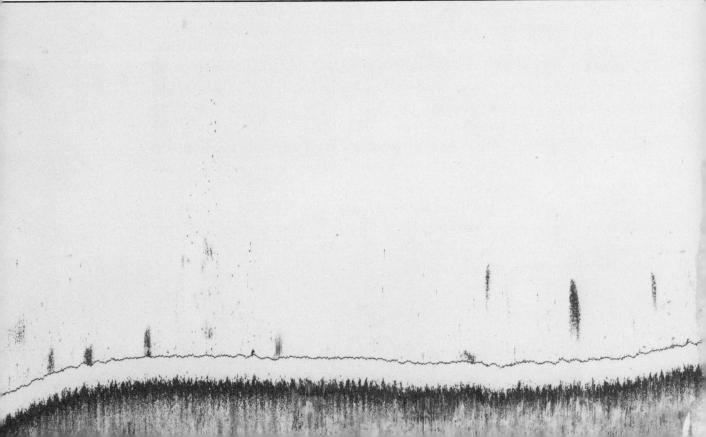

and the LORAN, he can feel his way over the bottom far below in much the same way that a man might fly an airplane above the clouds and still by instruments scan the ground. The *Elise G.*, under Kaare's control, prowls the ridges and gulleys looking for fish.

Eventually, whether it be minutes or hours, the dark blotches characteristic of schools of fish will begin to appear just above the depth sounder's tracing of the bottom. Kaare then doubles back, the net is put over, and a brief tow is made. Until the net is hauled back there is no way of knowing for sure what kind of fish are down there. It might be worthless dogfish that the depth sounder has picked up. It might be squid or butterfish, or it may be minnows so small that they pass through the net like magic.

If the school proves to be porgies, the real fishing begins. The

net goes back over and Kaare makes another sweep, twisting and turning to follow the sides of the ridges, pursuing the shadows of the fish as they register on the machine. Whenever he passes over a school he straightens the wheel to bring the net directly in line with the fish he has seen.

One successful tow may produce 50,000 pounds of fish. Often the net is so full that all of the catch cannot be taken aboard at once. When this happens the crew must perform an operation called splitting. A line called a splitting strap is tightened around the middle of the cod end. As the cod end is hoisted aboard by this line, part of the catch slips away into the wider belly of the net while the rest remains in the cod end. The cod end is brought aboard, emptied, and reclosed. The splitting strap is loosened and the cod end is put back over the side. Part of the belly of the net is then hauled aboard so that the remaining fish are once again packed tight into the cod end. This process may be repeated ten or twelve times in porgy fishing, while it rarely need be done even once when fishing for yellowtail.

A porgy trip may be over in two or three tows. However, the work of shoveling the mountains of porgies down into the hold, even if there is no trash to be picked out, will sometimes take hours. And after that, the real factor that determines the success of the trip will be the price for which the fish can be sold at the dock.

When porgies are for some reason scarce or if the weather has been bad enough to keep most vessels in port for several days, the price will rapidly climb. The fleet will sail with the first good weather and if fish can be found, the first few boats to return will be rewarded with a very good price. As more vessels return and more fish are landed, the price will steadily drop. In the space of a day the price may be reduced by half.

In this situation it very much favors a vessel to take a small quantity of fish and get them to port as quickly as possible. To stay out another day, even if this doubles the amount of fish in the hold, may, with a falling price ashore, result in less income.

A large powerful vessel such as the *Elise G.* is well suited to making the most of these price fluctuations. She can steam to and from the grounds at great speed, which allows her to go and return sooner than the rest of the fleet. Because she is better able to remain at sea and continue fishing in bad weather, it is often possible for her to come in with a load of fish when the rest of the fleet has had to remain in port for days. To land 100,000 pounds of fish when the price is up at 50¢ or 60¢ a pound is every captain's dream. With a vessel such as the *Elise G.*, with good fishing and good timing as well, this should be possible.

Another key to Kaare's success at porgy fishing is the amount of time he spends on the telephone while he is in port. He has friends on the docks and running boats in ports all up and down the coast. He talks to them frequently, gathering information as to how many vessels are out, when they left and when they are expected back in. Because the market price for porgies is set in New York, Philadelphia, and Baltimore, the prices paid for the fish where they are landed tends to be pretty much the same at all points on the coast. Thus, a large quantity of fish landed in Norfolk will cause the price everywhere to drop. Through his network of telephone contacts, Kaare is able to keep close tabs on the fishery.

Without the slightest twinkle in his eye, Kaare insists that whatever success he may enjoy in fishing has come by luck. In fact, there is little luck involved. To outwit gales, to keep a ship and crew operating efficiently, to find fish in the middle of a vast ocean all require more than luck. That the men who do these things well talk so much about luck may be in a small way the result of modesty, but it is mostly because the careful weighing of the many factors involved in each successful decision normally defies clear explanation.

Take, for example, a situation in which a captain has sailed with the intention of catching one type of fish but after several days is unable to find them. The principal choices are three: Stop fishing, return to port and take the loss; continue searching for what you have been so far unable to find; or try to find some other kind of fish to fill out the trip.

Considering just the last of the three choices brings up a host of other questions which must first be answered:

What other fish might actually be around; what have you heard on the docks recently, have you overheard anything on the radio; where were the various types of fish to be found last week, last year, and in the years before that?

Is the net which you have on board suitable for the new kind of fish; if not, can it be altered, is there spare material; how long will it take to do the work and, if necessary, how long will it take to undo it?

How far must the ship steam to get to the new fishing area; how long will this take and how much fuel will it cost?

What price may be expected if the fish are actually caught (this question requires for its answer a knowledge of the market standing of many different types of fish); how many other boats are fishing for that particular type of fish at this moment, and when are they likely to return to port?

Is there enough ice, fuel, water and food to permit more time at sea and, if not, is there a port nearby where these may be quickly obtained? What exactly will these items cost in that port and how much time and fuel will it take to get there?

How tired are the crew at this point in the trip? How tiring will be the new type of fishing; will it overtax them?

What is the weather likely to be? To extend the trip only to be stormbound and unable to fish would be a waste.

All of these factors and more must be assessed before a final choice can be made. And the thing that the fishermen call luck may result in no more than a gain of a cent or two per pound. It is true that once in a while there will be a big trip with a huge profit, but the real success in a lifetime of fishing comes from consistently making trips which land on the high side of a margin of a few cents.

And this is the craft of a shrewd and canny realist.

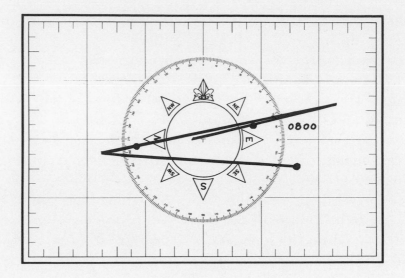

0800

The trash is now shoveled back over the side through openings along the rail, called scuppers. Each scupper is fitted with a steel cover which is kept closed until all of the valuable fish have been picked. The covers are then removed and everything goes out through them until the deck is absolutely clear of trash, sand, and debris.

The hose is then turned on the flounder in their pen. They are thoroughly washed with sea water, then separated into baskets; the large yellowtail into one basket, the small into another, all other

species into a third. John and Rob do this while Tom begins to clean the cod and haddock in the other pen.

Each of these fish must be cut open and the guts and gills removed before it may be stored in the hold. If not gutted, cod and haddock will quickly spoil. Tom works with great speed. It takes no more than eight seconds to gut each fish. In this tow they have caught only about sixty cod and haddock, so Tom is finished gutting all of them in about ten minutes.

Many more cod and haddock are often caught. Some trips the entire crew will gut fish from 6 in the morning until well after midnight, only stopping now and then to haul back the net and dump another pile of fish on the deck.

Gutting is the sort of work that sends a person into a trance. It is fast and messy. Speed depends on a mindless quickness of hands. Tom arranges everything carefully before he begins. He lays a wooden crate on its side and slips a plastic bucket under the edge of it at one end. He sits at the other end of the crate facing the bucket and arranges the fish on the deck so that he may easily reach down with his left hand and pull each fish onto the crate in front of him. He holds the knife in his right hand and chops down on the throat of each fish. With the same motion he rips open the belly; then with middle and index fingers of the knife hand he hooks the arches of the gills and rakes back down the length of the body cavity. He slips the handful of guts into the plastic pail and the empty fish onto the deck to his right.

This morning the stomach of each cod contains one small fish called a hake. Each of the hake is 5 inches long and has been equally digested by each cod. It appears that at some time around dawn every cod in this area seized and ate one hake. At different times of

the day most species of fish swim at different depths. At night, for instance, cod swim nearer the surface than in daytime, when they stick mostly to the bottom. It is possible to imagine schools of fish swimming at different depths in broad, thin layers. Between night and day these layers of fish pass through one another. This morning a layer of cod passed briefly and ravenously through a layer of five-inch hake. It is possible only to guess at the forces that gather, shift, and combine far beneath the surface.

Cod

GADUS MORRHUA

Europeans probably fished for cod off the coast of North America before Columbus made his southern discovery in 1492. But while the news brought back by Columbus caused whirls of activity all over Europe, the discoveries of the Newfoundland fishermen were hardly noticed. At that time people had plenty of food but not enough gold. The small amount of treasure brought back by Columbus's ships on his second voyage excited every king in Europe. The steady landings of dried cod excited no one at all.

The fishermen of this time came and went as they pleased—gone for whole seasons at a time. It didn't matter where they went to fish so long as they came back with their fishholds full to the hatches. The fishermen sailed regularly to an unknown land, not drawn on any map. When they came back they spoke to their friends of Indians, forests, and white bears, but no one cared. The fish were landed, eaten; and the boats went out again.

Codfish could be found in great quantity. Each fish seemed eager to bite the hook; they could be hauled aboard almost as fast as a shepherd drives sheep through a gate. They were a perfect food for a civilization about to invent industry, politics, and theories of money. Codfish could be salted, later dried, and made to keep indefinitely.

Packed in barrels of brine, split cod traveled everywhere. They went by wagon far inland, by mule over mountains, by ship across the sea.

Fishermen once believed that they could foretell a coming storm by examination of a codfish's entrails. The fish swallowed stones, they said, to hold them to the bottom.

Long ago when cod were cleaned for salting, the head would be removed by one man and the body passed on to another man for the next step of the cleaning process. If the man receiving the body found that it still twitched with nervous life (as freshly killed fish often do), he would demand that the man who had severed the head go back and strike the loose head a sharp blow with the blunt handle of his knife. This treatment would in every case immediately stop the body's twitching.

Because it will eat almost anything and because it travels through a great variety of depth and undersea terrain, the cod was once an instrument of science. Early marine biologists made a practice of examining the contents of cod stomachs for examples of as yet undiscovered marine species. Many sea creatures bear as their Latin names the name of the scientist who first discovered them in this way. If credit were given where credit is due, these discoveries would bear the name of *Gadus*, the codfish that brought them to light.

Today, cod grows ever more valuable and ever more scarce. No longer salted or dried, it is sold either frozen or fresh. In many cases it is processed by machine. Codfish has been the subject of intense debate among the United Nations. Recently, cod was the cause of a small war between Iceland and England.

But still cod swim in the sea just as they did at the time of Columbus. They are no more aware of their importance today than they were then. And still the fishermen leave port in search of them. They go out in small ships, alone into little-known lands. After a week or a season they return with holds filled to the hatches. But no one on shore takes time to ask them where they have been.

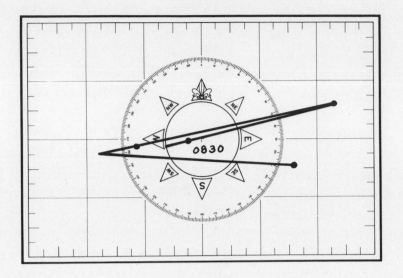

0830

Once all the cod and haddock are gutted, they are washed down with the hose. Then they are sorted into wire baskets. The two species of fish are kept separate in the hold because together they will spoil much faster than apart.

The baskets are counted and the number of each species is recorded on the tally sheet that Kaare keeps in the wheelhouse. This tow there are twelve baskets of large yellowtail, one of small; two baskets of cod, and one of haddock. Tom then slides open the hatchway leading to the fishhold and climbs down.

The fishhold is wreathed with a frosty mist. The air is heavy and chill; there is a slightly sour smell from the thousands of fish that lie there buried in ice. The hold measures about 25 feet long, 15 feet wide, and 7 feet high. It is divided in half by an alleyway. On either side of the alley are rows of stalls, called houses.

Fish are poured directly into each house through small manhole openings in the work deck. Each manhole, or "deck plate" as it is called, is securely closed by means of heavy, watertight latches so there is no chance of water leaking into the hold. A small amount of water will spoil the fish. Large amounts will sink the ship.

Before any fish are put into a house, Tom must first put down a bed of ice. Any fish that touches the floor or walls of the house will soon spoil, so the ice must be about 6 inches thick to be sure that it will not melt away before the fish are taken out in port.

The front of each house may be closed off by stacking wooden boards on edge across the opening. As each basket of fish is poured through the deck plate, Tom sweeps a shovelful of ice over it. As the fish pile up in each house, Tom puts more boards across the front of the house until it is entirely filled. Each house will hold about 5,500 pounds of fish plus the ice necessary to keep the fish fresh.

Tom must see to it that each fish has enough ice to last the trip home. Especially during summer the ice melts rapidly. Once the ice is gone the fish will spoil in a few hours. But Tom must also be sure not to use too much ice or he will run out before they have caught a full trip. It requires about 18 tons of ice to make a summer trip to Southeast Part.

Tom must also pay attention to the "trim" of the vessel. That is, he must be careful not to stow too many fish on one side or the other of the hold or else the *Elise G.* will develop a list—begin to tip to one

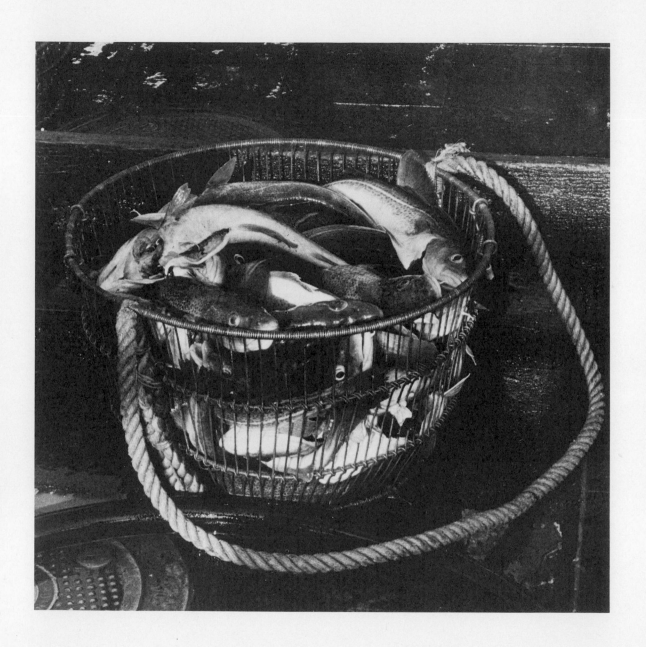

side. Although severe storms are uncommon at this time of the year, there is always danger if the boat is out of trim. In the worst conditions the extra few inches caused by careless placement of fish in the hold could make the difference between the ship's rolling over and staying afloat.

Kaare has built a reputation for landing only good quality fish. He is very careful of that reputation. He uses a lot of ice on the fish in the hold. And he stays out for shorter periods of time than many other fishermen so that the fish spend less time in the hold. He insists that the fish be entirely free of sand and other foreign matter before they are put below. He makes shorter tows of the net than most captains so that the fish are less crushed and net wrung. Fish dealers are glad to buy Kaare's fish because they know that it will be of fine quality. In this way Kaare knows that he will always get top price. Some captains are not so careful.

The largest part of the cost of each trip comes from the fuel burned getting to and from port. Because of this, many captains prefer to stay on the fishing grounds for as long as possible before going in. They feel that they should catch as much fish as they can before having to face the expensive run home. Many of these captains make trips of twelve and fourteen days, as opposed to the seven- or eight-day trips made by the *Elise G*. Fish that have been in the hold for twelve days are not as good as fish that have been in the hold only six. Particularly in summer, when the ice between the layers of fish melts rapidly, the oldest part of the catch from a long trip is often spoiled by the time it reaches the dock. The captains of these vessels know that a part of their catch may have to be sold at a reduced price or even dumped.

Lumpers, Sorters,
Buyers, Truckers, Cutters,
Wholesalers, Retailers

When it comes time to unload, the fish and ice are removed by taking down the boards across the front of each house and pitchforking the fish into baskets. The baskets are then hoisted up through the large hatchway and swung up to the sorting platform on the dock. The actual unloading of the houses is usually not done by the crew but by workers hired from the fish dock. These men are called lumpers and their job is to "lump the fish out."

As the lumpers heave away at the masses of ice and fish, Tom operates the small winch which hoists each basket out of the hold and onto the dock. The controls for this winch are located in the wheelhouse at one of the rear windows overlooking the work deck. The control is a single lever which hoists the basket when pushed forward, lowers the basket when pulled back.

John stands on the deck beside the hatch. Each time the lumper hooks the hoisting line onto a basket, John gives a thumbs-up signal to Tom. As the basket comes up through the hatch, John takes hold of it and smoothly swings it toward the sorting platform on the dock 20 feet away and 10 feet above the rail of the *Elise G*.

Tom continues to hoist as the basket swings. Rob, who is standing next to the platform, catches the basket and slides it onto the platform. Tom reverses the winch to give a little slack in the line and Rob overturns the basket, dumping the fish into the long, stainless-steel tray of the platform. He then swings the empty basket back to John, Tom lowers away, John catches it and guides it back down through the hatch to the lumper.

The lumper always has several baskets in the hold so that whenever the hook returns with an empty basket he is ready to hook a full one on and the process is repeated.

At the platform, Rob and Kaare sort the fish along with several fish-dock employees. The yellowtail have already been separated into small and large size, so the sorting amounts only to removing chunks of ice and a few bits of foreign matter before the fish are spilled off the end of the platform into ice-lined crates, each designed to hold 125 pounds of flounder.

The size of the flounder, and thus the price that will be paid for them, is determined by the "count"—the number of fish required to make up the total weight of one 125-pound box. If it takes fewer than 120 fish, the box will be sold as large yellowtail for a high price. If it

takes more than 120 to fill the box, the flounder must be sold as small for a lower price. The count is taken several times during the unloading process at the discretion of the buyer.

As each crate is filled, a shovelful of ice is put over the fish and a top nailed on. The crate is then marked and stacked with the others from the *Elise G.* along the wall next to the sorting platform.

If the unloading is begun at 8:00 in the morning, it is usually finished by 2:00 in the afternoon. Once all of the fish are out, the total amount of the catch is tallied and its value determined according to the market value of the various species of fish.

The *Elise G.* and her crew receive their money from the buyer and immediately begin preparing for their next trip.

The yellowtail flounder that the *Elise G.* is catching this Sunday morning on Southeast Part will be lumped out in Newport on Tuesday morning. From there it will be shipped by truck to New Bedford, Massachusetts, 25 miles away. There, it will be bought by a cutting house where it will be kept in cold storage until Thursday morning, at which time it will be filleted (each fish cut into two boneless strips of meat) and packed with other fillets in 20-pound tins. That afternoon the fillets will be put on a truck bound for New York. There it will be unloaded before dawn on Friday morning and sold to a restaurant or fish market. By this time the *Elise G.* will be back out on Georges Bank beginning the first tow of another trip. If all goes well the fish will be eaten Friday evening, six days after it was caught and still tasty enough to be called fresh.

It is possible, however, that the fish will be delayed at some point along its route. It might remain an extra day or two in the buyer's or the cutter's cold-storage. It will likely be held up over at least one weekend.

The fish will be briefly held by as many as eight people before

it is eaten. Kaare and the crew of the *Elise G.* have possession first. They sell it to the buyer at the fish dock. He may sell it to the trucker who takes it to New Bedford, and there it is sold to the cutting house. It then goes to another trucker who takes the fish to New York, then to a dealer and perhaps still another trucker, then it goes to the restaurant- or market-owner. Finally, the customer takes possession of the fish, carries it home, and eats it.

Of course, each time the fish changes hands, its price goes up. On Tuesday morning, if the market is good, Kaare will be paid 50¢ a pound for the large yellowtail and 35¢ a pound for the small. Friday afternoon, in a small fish shop in New York City, the very same fish will be sold at $3.05 a pound.

Nowhere along this complex chain of price increases does anyone make an unreasonable amount of money. Each person in the long chain can show you that his profit is but a small one, barely enough to keep him in business, he says, with times as hard as they are today.

It is also a fact that Kaare and the crew make the best percentage of money from the chain. With a good boat and with hard work, it is possible to make a very good living at fishing today. In the past, this has not always been the case. There have been periods when fish were too scarce to depend on, and even though prices tend to be high at these times, not enough could be caught to make a good living. At other times, fish were too plentiful, and the prices would drop to pennies a pound. There have been times when the whole marketing system was heavily weighted to favor the various middlemen rather than the fishermen. But right now things are good. There is money to be made by going to sea.

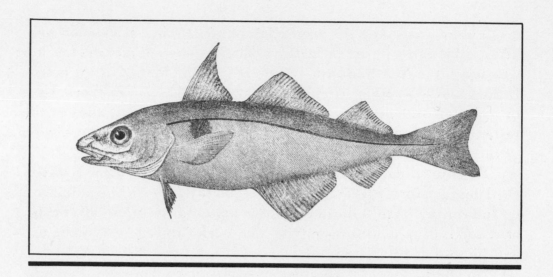

Haddock

MELANOGRAMMUS AEGLEFINUS

Haddock normally weigh 3 to 4 pounds. They may reach 25 pounds and 3 feet in length, but this is very rare. They frequent water from 30 to 1,400 feet deep, ranging in temperature from 1° to 10° Centigrade. They reach market size at the age of three years and may live fourteen years. They feed on the same range of things as do codfish, their first cousins. Haddock look very much like cod in shape but are silver-gray with a small dark patch just behind the head and a thin black line running the length of the side.

In late winter, haddock gather in dense schools to spawn on the easternmost shoal of Georges Bank. This area is known as the Winter Fishing Ground, for wherever the haddock go, the fishermen follow. Haddock were for almost a century the most sought-after fish in New England. They could be easily caught and kept fresh with ice. Their flesh was white and firm, and enjoyed a good price ashore. On a good day, a schooner's crew could catch 25,000 pounds of haddock, 70,000 or 100,000 pounds in a trip. When the fishing was good, it was very good.

But even in the early days, before man caught more than a small proportion of the haddock that swam on Georges, there were alarming fluctuations in the species's numbers. Old accounts indicate that haddock could not be found on Georges in the 1850s.

These natural fluctuations are apparently caused by the haddock's inability to spawn successfully in some years. Haddock eggs, though they are laid near the sea bottom, soon float to the surface. If the water temperature is 4° Centigrade, the eggs will float for twenty days before hatching. In water seven degrees warmer, the process will take ten days. In water either hotter or colder than this range, the eggs are likely not to hatch at all.

The strong winds and currents on Georges as well as the nearby masses of warm and cold water (warm in the Gulf Stream, cold in the waters off the coast of Maine) can cause great temperature differences from one year to the next. If conditions should be wrong for several years in a row, the haddock become very few; several good years in a row and the population will boom.

Scientists have for many years kept track of the abundance of haddock. Each year they try to determine if the haddock have successfully spawned the year before. To do this they sample the fishing grounds using fine mesh nets capable of catching young fish that will not be caught in the fisherman's nets for another two years. By counting the number of one-year-old fish, the scientists can get a good idea of how many fish will reach marketable size two years later.

If the scientists find few young fish one year, they will put a limit on the number of fish that may be caught when that generation of fish reaches market size. By keeping careful watch over the number of haddock, the scientists believe that Georges could be made to produce 55,000 tons of haddock each year.

In 1962 and 1963, two very successful spawning seasons occurred. Scientists predicted that great schools of haddock would reach market size in 1965 and 1966. Everybody got ready.

The foreign fleets in particular made plans to concentrate on haddock. Soviet ships were sent in force. In those two years over 200,000 tons were caught—twice the safe amount! At the same time, the haddock suffered several bad breeding seasons, probably because of the intense disruption of the spawning grounds by the hundreds of vessels towing back and forth across them.

In the space of a few more years, haddock were almost extinct on Georges Bank. In 1970, '71, and '72 the limit was set at 13,000 tons per year. Today the yearly catch of haddock is limited to half that figure—one-tenth of the number taken annually before 1965. Also, the government has called for the closing of the entire eastern portion of Georges Bank during the haddock's breeding season.

It is too early to tell if these strict regulations will do anything to save the haddock from commercial extinction.

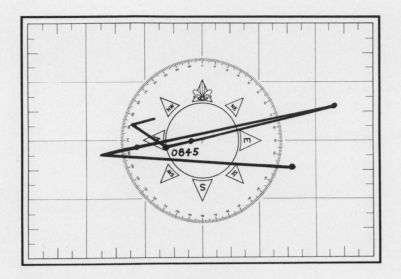

0845

Suddenly Kaare signals for the net to be hauled back. Tom spreads a last shovelful of ice on the fish in the hold and hurries back on deck. The net has been on the bottom for only forty minutes, too little time to have caught much. Kaare must suspect that something is wrong.

The *Elise G.* tows with great force. Anything that may snag the net will certainly tear it. Kaare is always alert for any change in the ship's towing speed or direction, any difference in the vibration of the deck that might signal the net being damaged. If the net snags

only in the twine, it may rip away with nothing felt at the surface. But if the wires or chain become snagged, there will be ample notice in the wheelhouse. The vessel may stop dead in the water, veer sharply to one side, or there will be a shudder which runs up the wires and into the steel of the ship like a chill.

The net comes to the surface with its belly torn completely away. There is only a small quantity of fish in the cod end, seven baskets of yellowtail. These are dumped out on deck, the *Elise G.* is allowed to drift, and everyone sets about repairing the net.

An inexperienced person would not even know where to begin. The net lies spread on deck in great tatters; whole sections are missing entirely. Kaare picks up one corner of the net and begins counting hand over hand each mesh along one side of the missing belly. The tear is irregular in shape, so that from time to time Kaare must count several meshes across, then resume counting down the side. He memorizes the shape of the hole by the counting of the mesh. Kaare sends John to the bow of the ship to fetch a spare piece of netting with which they will replace most of the missing parts. He sets Rob to cutting away those tatters too raggedly torn to be easily restitched. The objective is to first reduce the damage to a regularly shaped hole into which can be fitted a new section of netting. The seams and places where the net is only ripped with no twine actually missing can then be quickly rejoined.

The work must be done very precisely because when fishing every part of the net is under strain. If there is a tight spot caused by a repaired section being too small, that spot will immediately tear out again. If the repair is made too loose, it will put extra strain on the rest of the net and cause tearing elsewhere.

It requires great skill and experience to mend quickly and well. Kaare and Tom are both good at it and work on the most difficult parts. Rob and John take the simpler tasks but they are expected to learn by paying careful attention to what Kaare and Tom are doing. Careful observation as the work goes on is the only way to learn to mend for no one has spare time enough to give a lesson. When the net is torn up and lying on deck, everyone on board is losing money. The most important thing is to get the net back on the bottom.

It is two hours—almost 11:00—before the net is finished. Hoping to avoid whatever tore the net, Kaare moves about a mile to the north for the third tow of the day; closer to the other boats.

Towing and Mending

The cautious fisherman will collect the location of each obstruction—wreck, rock, ridge, and snag—and mark it on his chart. He will seek information from other fishermen; he will learn from them the location of more and more spots that he must avoid.

Over a lifetime his chart will grow increasingly freckled with danger spots and the places the man feels safe to set his net grow few. Kaare says there are only a small number of places entirely free from obstructions and these, because they are safe are also the most heavily fished. Heavy fishing leaves few fish uncaught, and so where the fishing is best from the standpoint of protecting the net, it is probably worst from the standpoint of catching fish.

Kaare keeps no charts of obstructions. "No," he says, "you got to go and fish just where you think the fish are, and if you tear up, well, then you mend."

There is the story of two brothers who fish off the Rhode Island coast, mostly in a place called the Southwest Hole. One arm of this place is so filled with boulders and snags that it was considered impossible to fish. But the two brothers went back again and again. More often than not they would tear up and lose gear, but once in a

while they got through safely and each time they made a clean tow they carefully wrote down exactly how they had done it. They towed and mended, mended and towed as they explored the labyrinth of safe paths through that part of the Southwest Hole. One by one they discovered the alleys along which their nets might safely be drawn. The fish were plentiful there, and the brothers have recently enjoyed great profit. In the rest of the Hole where the rest of the Rhode Island fleet works, the catches have become smaller and smaller. Those without knowledge of the rocky ground are discouraged and have begun to move away.

Likewise, on Georges Bank there are many vessels that make a specialty out of fishing rough bottom. Far up among the ridges and gulleys of the actual shoals there are plenty of cod and haddock, but the likelihood of tearing up is very great.

The boats that fish there are specially rigged. Their nets are fitted along the sweep with large steel or rubber rollers so that the net will ride up and over the many rough spots. They carry extra men and all are expert at mending. One crew fishing on rough bottom reported that they put thirteen new bellies into their net. They said they were able to put in an entire belly in just twenty minutes, and that using only two men! These boats are rigged with two nets, one on each side of the boat, so that when one comes up torn the other may be immediately set in its place. In this way the vessel is able to keep right on fishing while repairs are being made.

Net repair is the single most important skill that a fisherman must learn. If he is skillful, he need fear no obstruction and may fish wherever it pleases him.

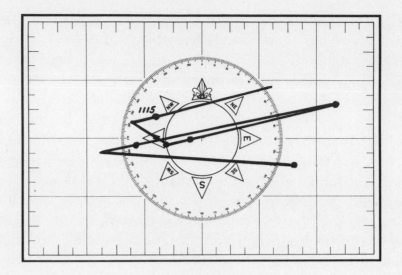

1115

Just inside the heavy steel watertight door that leads in from the work deck, is a wooden door and behind it the steep ladderway that descends to the engine room. The door is kept tightly shut.

The *Elise G.* is steaming westward at full throttle to a position closer to the other vessels. The engine is heard in two ways: as a deep rumbling that rises up through the steel hull and deck of the ship like a feverish earth tremor, and as a high pitched, distinctly musical tone that bellows with a full throat of black smoke from the ship's tall twin exhaust stacks.

Twin trumpets, the stacks produce a differing note depending on the setting of the throttle. The slower the engine is turning, the less the volume of exhaust and the lower the note. As the throttle increases, the pitch of the stacks, and the vibration of the entire ship, increase also.

It is the musical element of the *Elise G.*'s exhaust that can be heard from the greatest distance. The pure note carries infinitely farther than the roaring and rumbling that make up the balance of the whole engine noise.

John opens the engine room door and descends to get a dry pair of gloves from a clothesline that is strung above the engine. The heat, the smell of hot diesel and lube oil, and the sounds that rush together up the ladderway and out through the open door are astonishing. Walking against this sensory current, John starts down the ladder.

Compared to the pleasant living quarters and the open, uncluttered work deck, the engine room is a stark and threatening place. The decks are of gray-painted steel. The sides and the overhead are a plain white. In the middle of the space, bolted securely to heavy steel beds welded in turn to the floors of the hull itself, is the main engine. It is a General Motors 12-cylinder diesel, Model Number V-149, designed to produce 675 horsepower when the throttle is set at 1800 rpm. To do this it must burn 28 gallons of fuel per hour and consume tens of thousands of cubic feet of air. It is the final mixture of this fuel and air, exiting through the exhaust stacks, that makes the fine, far-carrying trumpet note. It is the energy from this mixture transmitted through a massive steel shaft to the ship's propellor that drives the *Elise G.* ahead.

The engine is painted green. Except for the propellor shaft and one or two belts that drive auxiliary gear such as the alternator and a pump, there are no visible moving parts. It is startling to stand before the engine as it gives off its ear-splitting howl and to be able to see not the least sign of its working.

Mounted just forward of the main engine are two 20-kilowatt electrical generators, each powered by a General Motors 2-cylinder diesel engine. Only one is running and because it is much smaller it cannot be distinctly heard against the main engine. However, if it were to change speed or stop altogether, there would be an obvious change in the vibration of the whole vessel.

And everywhere there are pipes, hoses, and wires, all neatly fitted to pass alongside, around, or across one another as they rise from their individual sources and lead each to its private purpose. Everything is arranged in an orderly way, the pipes and wires of different systems painted different colors so they may be followed through the fabric of the ship.

Mounted on the front of the engine is the hydraulic oil pump which, by way of steel pipe and high pressure hose, delivers fluid power to the towing winches, to the net drum, and to the small hoist used for taking the fish out of the hold in port. Once used, the oil recirculates by way of still more pipe and hose to a large reservoir where it cools, is filtered, and soon is drawn back through the pump once again.

There are riddles of valving and pipe that deliver cold sea water to cool the main engine and generators. There are banks of pipe and valves to and from the fuel tanks which together hold 14,000 gallons, enough for the *Elise G.* to steam nonstop for 5,000 miles.

There is a pump and plumbing system to transmit fresh water from a 3,000-gallon tank to the living quarters above.

There is a maze of wiring and circuit breakers to power the many electrical systems of the vessel. There are two large emergency pumps, one belt-driven from the main engine, and the other electrically powered. One of these pumps is normally used to draw in sea water to be discharged by way of the hose on deck for washing down the catch.

The lighting is bright. It is supplied by bare bulbs enclosed in explosion-proof glass-and-bronze fixtures. The ventilation of the space is supplied by a powerful blower which pulls air through a large steel intake that opens seven feet above the work deck and is fitted with a cover designed to admit air but at the same time prevent water from entering in the event of heavy weather.

John has just come to get a pair of gloves. The noise is overpowering, violent. There is a strong urge to run from the sound, a fear that some permanent change is being done to any that may stand near for too long.

John plucks the gloves from the line and hangs the wet ones in their place. He moves, with the caution of a man in the dark, back to the ladder; up it and out the door; closes the door against the heat and the infernal din.

"Pheeew! That's some kind of place," he says.

Wrecks

Everywhere on the continental shelf lie sunken ships. Near the *Elise G.* today are the bones of the *Seiner*, a trawler last seen on January 13, 1929. In the weeks that followed her disappearance with all of her twenty men, other trawlers dragged up her broken lifeboats and bits of her rigging from the bottom. No one knows what disaster befell the ship, but whatever it was, it must have been swift. Twenty-five miles to the northeast lies the *William H. Starbuck*, torpedoed by a German U-boat on August 10, 1918. Collision, fire, explosion, and storm have claimed hundreds of ships on Georges Bank.

By far the greatest number of wrecks on the fishing grounds are those of fishing vessels themselves. The first fishermen to visit Georges Bank sailed from Gloucester, Massachusetts, before 1830. They didn't like what they saw. The tidal current seemed to flow too strongly for them to anchor safely. They feared that in a heavy sea the current would pull their small vessels under. Also, the waters were shallow and the frequency of easterly gales made the entire area treacherous.

Before leaving Georges, they lowered their baited hooks to the bottom and found abundant halibut, cod, and haddock. But was the fishing worth the risk? They thought not, and so sailed home.

But not long after this first visit the demand for fish began to increase. The schooners, by now a little larger, a little more sea-worthy, returned. They anchored, they fished, and they made money.

Soon, the early practice of fishing from the side of the ship was replaced by a technique in which the entire crew except for the captain and the cook would daily row out from the ship alone in small boats called dories. A man would tend two or three handlines, each with a single baited hook and a 2½-to-8 pound lead weight to carry it to the bottom.

Still later it became the practice for two men to go in a dory and together tend one line better than 2 miles long with up to 3,000 hooks spaced along its length. This line, in those days called a trawl and today called simply a long line, was stretched out along the bottom, allowed to "fish" for several hours or else overnight, and then was pulled back into the dory. The fish were removed and they and the trawl were rowed back to the schooner, the fish to be cleaned and either iced or salted, the trawl to be rebaited and coiled down for the next time it was to be set.

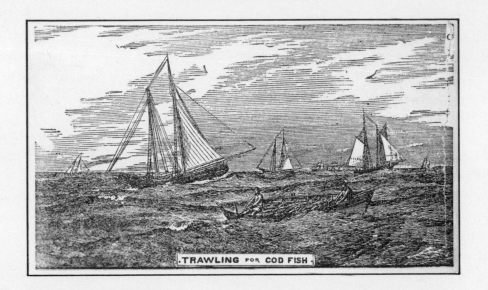

.TRAWLING ᴘᴏʀ COD FISH

The best fishing was for haddock and occurred in late winter and spring. The best place to fish was on the northeast edge of Georges Shoal, the Winter Fishing Ground, close to shallow water. The winter gales which sweep this area blow from the deep sea toward the shallows, driving tremendous seas over the gravel bars of the shoal. Any vessel unfortunate enough to be blown into these breakers would be bludgeoned to pieces as each wave alternately lifts it, then dashes it on the bottom.

To escape this graveyard, the fishing vessels nearby would try either to sail for open water or else remain at anchor and hope to ride out the storm. Either choice had its fearful side. To sail off meant driving hard into the teeth of the gale. The beating a schooner would take under these conditions could be terrible—easily enough to disable it and send it drifting back to certain destruction. To remain at anchor would also mean a frightful hammering. As they clung to their anchor cables, vessels were stripped of railings, deckhouses, masts, dories, sails, and men. Many sank while others broke free and were cast into the shallows. Often, when many schooners were anchored in the same area, one vessel, on breaking loose, would drift down upon several more, breaking them loose in turn or else sinking them outright.

In blizzard, when the fleet was anchored tight together on the Winter Fishing Ground, the entire crew would remain on deck, one

man stationed with an ax in the bow, others ready at a moment's notice to hoist sail. To cut the anchor cable and attempt to dodge another vessel, then to be forced to sail offshore through the anchored fleet, was the last and the deadliest of chances. It was not uncommon for the crew of one schooner to watch helpless while several vessels around them, each containing better than twenty men, went down.

In years when gales were frequent, the loss of vessels and lives was staggering. In the nine years between 1874 and 1882, the port of Gloucester alone lost more than 120 vessels and 950 lives.

Many deaths also resulted from the loss of individual dories while they were away from the schooner. If the weather should suddenly turn bad or the visibility fail due to fog or snow, the men could easily become lost and the schooner unable to find them. A slight mistake in handling could easily overturn a dory. Few fishermen ever learned to swim very well.

Modern fishing vessels operate in much greater safety. They have engines powerful enough to drive them through almost any weather. They have radar and LORAN. They can pick up weather forecasts to warn of an approaching storm long before they would be able to predict it from the entrails of a cod. But even in the 1970s, several vessels are lost each year. Because of two-way radio, fast rescue ships, and helicopters which can pick up survivors far at sea, these sinkings usually occur without loss of life. But still, vessels, usually those worked years past the time they should have been safely retired, one by one settle to the bottom.

By marking the location of wrecks and boulders on the chart according to their LORAN positions and by then being sure not to tow across those spots, a fisherman tries to keep his nets from being torn apart. But every now and then a new wreck is found or someone makes a mistake and part or all of a net is snagged. Every year thousands of square yards of netting are lost. Divers report that many of the wrecks on the grounds are covered with pieces of these lost nets.

The pieces hang suspended in the water around the wreck, and as they hang there they continue to catch fish which by accident swim into them, become tangled, and die.

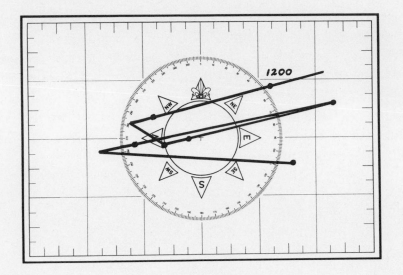

1200

The *Elise G.* passes close to another vessel. It is the *New Land* from New Bedford, Massachusetts. She is an eastern rigged dragger, a side trawler typical of the hundreds of fishing vessels built from 1935 to the present. She was made of wood in 1965 at Thomaston, Maine, and has fished out of New Bedford ever since.

The most noticeable thing about the *New Land* in comparison to the *Elise G.* is that her wheelhouse is located as far back in the stern as possible. In front of the wheelhouse is placed the work deck. Along the rail on each side is located a pair of gallows. Like most

New Bedford boats, the *New Land* is rigged to fish from either side.

When fishing, the tow wires both run out over the same side, one from the gallows mounted in the bow and the other from the gallows mounted back next to the wheelhouse. There is no stern ramp, so the net must be set and taken aboard over the side between the gallows.

The towing winch on the *New Land* is mounted directly in front of the wheelhouse. From it, the two wires run forward along the deck, through massive pulleys, called bollards, set in the deck, then out over the side through the towing blocks set, as in the *Elise G.*, at the peak of the gallows. The pens for sorting fish and the hatches leading down into the fishhold are placed between the wires and bollards on deck. There is far less open space. There are many more objects to trip over and slide against. The wires, arranged as they are to run at ankle height through the middle of everything, are a constant nuisance and danger. But the crew, working long days with the wires underfoot, pays little attention to them.

The *New Land* is several feet shorter than the *Elise G.* and has only half the horsepower. The net she tows is only about half the size of that handled by the *Elise G.*

But the greatest difference between the two vessels is the way in which the net must be handled. Setting and hauling back over the side requires that the *New Land* be completely stopped much of the time. Furthermore, it means that the net must always be handled on the windward side of the vessel. This is done so that the boat will be constantly pushed by the wind away from the net lying in the water alongside. If the boat were to drift over the net it would become hopelessly tangled.

Every time the *New Land* hauls back, she must steer in a wide circle so that the wind and net will be on the proper side. She must

then lie dead in the water while the cod end is brought aboard. Because she is not equipped with a power drum to wind up the net, all of this work must be done by hand. When the cod end is close alongside, it is attached to a block and tackle and hoisted clear up the side, swung in over the deck, and let down. All this added labor means that the *New Land* must carry at least one more crewman than the *Elise G.* An extra man means greater expense and a smaller share of the profits for everyone.

Lying with one side to the wind means that that side will also be exposed to the full force of the waves. A side trawler will roll deeply when the net is alongside. Under these conditions, hoisting the ponderous cod end into the air and then lowering it onto the deck is a very tricky matter. The crew must exercise perfect judgment and timing to handle the net and at the same time avoid being crushed by the full cod end. If the sea grows increasingly rough, the side trawler must stop fishing altogether.

The *Elise G.*, a stern trawler, faces none of these difficulties. Because the net is handled entirely over the stern, it is possible to keep the hull from becoming entangled in the net by steaming slowly ahead, thus causing the net to stream out neatly behind. Because it can use its engine to position the net when setting and hauling back, a stern trawler need not go through any complex turns or stops. This saves a great deal of time, which in a full day's fishing will add up to time enough for at least one complete tow.

If it gets rough, the stern trawler may set and haul back while heading into the seas. In this way the vessel rolls far less. The raised deck and wheelhouse in the forward part of the vessel provide a wind break and shelter for the men on the work deck. Because of the stern ramp, the cod end can be taken aboard without being lifted into the air and given the opportunity to swing and hurt anyone. The design of the stern ramp is perfectly suited to the use of a motor-powered net drum so that none of the net handling need be done by hand.

The fact that stern trawlers seem to do everything better than side trawlers makes one immediately ask why stern trawlers weren't developed decades ago, as soon as there were engines strong enough to pull a net. The answer is that American fishermen have always been very cautious about adopting new methods. By 1900 the otter trawl and a similar type of drag net called the beam trawl had been in wide use in England and Europe for many years. A beam-and-otter-trawl fishery had also existed along the shore of Cape Cod, Massachusetts, for almost twenty years. Clearly it was time for American fishermen to try the otter trawl offshore.

But the big event in the New England fisheries of 1900 was the launching of the schooner *Helen Miller Gould.* She was a purely traditional sailing vessel except she was fitted with a small gasoline engine to push her about should the wind fail. To the Gloucester fishermen she seemed a most daring experiment. A little too daring perhaps, because eighteen months later her engine went afire and she sank.

Finally, in 1905, a group of Boston fish dealers banded together to build the *Spray,* an otter trawler copied from the big steam-powered trawlers common at that time in England. By 1914, there were a dozen big otter trawlers operating out of Boston and New York and the superiority of the type had been clearly proven.

But still the old sailing vessels kept on. As late as 1954, there was still one dory fishing schooner operating under power out of Boston and Gloucester. The rest of the old schooners were one by one converted over to otter trawling by the addition of large diesel engines, winches, and gallows. This period of conversion of old schooners from dory fishing to otter trawling spanned nearly twenty years.

And even as new vessels were built especially for otter trawling they retained much of the appearance of the earlier schooners. Thus the *New Land,* fishing this morning alongside the *Elise G.,* is not much different from the converted schooners that came before her. The placement of her wheelhouse far back in the stern, the position of her mast, and the modeling of her hull all suggest her New England heritage.

The appearance of the *Elise G.* suggests an altogether different evolution. The thinking that created her began in Europe in the 1950s. With the development of relatively inexpensive welded-steel construction it became possible to design a ramp through the stern of a fishing vessel. The advantages to this system seemed immediately obvious to everyone.

But most American fishermen who heard of these plans warned that although it sounded *theoretically* possible to operate such a vessel, *practically* it wasn't worth it. A reasonably priced, safe vessel just couldn't be built that way.

By the late 1950s, big European stern trawlers began to appear on Georges Bank, then Canadian stern trawlers arrived, and finally in 1964 two U.S. stern trawlers were built.

But just as it took almost twenty years for the otter trawl to prove itself over the hopelessly dangerous and inefficient method of dory trawling, so it took almost as long for the stern trawler to win acceptance over the old-style wooden dragger.

It was not until the 1970s that American shipbuilders in the South began to get orders for stern trawlers. Even today there are still a few side trawlers being built in New England and there is still debate as to whether or not vessels like the *Elise G.* are too expensive, too complicated, or not adequately seaworthy to fish off New England.

Kaare Gjertsen has had ample opportunity to consider all sides of the problem. He was born in Norway in 1925. He became a fisherman at age fourteen when he signed on board a fishing vessel in the North Sea. He moved to the United States in 1948 and continued fishing out of New York and New Jersey. In 1954, he had saved enough money and gained a good enough reputation as a fisherman to buy his own vessel, the *Mary M.*, a side trawler built in Maine.

Eight years later, he sold the *Mary M.* and bought a newer Virginia-built boat which he named the *Nettie R.* For fourteen years he kept her fishing hard, all the while experimenting with different otter-trawl designs and different ways of handling the trawl on deck.

Then Kaare began to think of building a new boat. He wanted speed to enable him to change fishing grounds easily—to hunt, and to get to and from port quickly. He wanted a vessel easy to work with few men. He wanted a boat that would be comfortable and safe to work on. He wanted a vessel that could be used in a number of different types of fishing. And above all, he wanted a vessel that would catch lots of fish.

In the spring of 1976, Kaare took a trip south to visit different shipyards. In Alabama, he found the builder he was looking for. Kaare selected the general appearance and hull shape for the new boat out of standard designs developed by the shipyard. He drew the engine-room and work-deck layouts that he wanted. Kaare then negotiated a loan from a bank in his home town. With his great experience as a fisherman and with the money that he had saved over the years, he had little difficulty in getting the extra money he needed.

The new vessel was launched in October 1976 and has fished hard and profitably since. The boat represents the sum of Kaare Gjertsen's long and successful career at sea. It is the product of a life of hard work.

Kaare named the vessel after Elise Gjertsen, his wife.

The Foreign Fleet

Although she represents the most modern type of U.S. fishing vessel, the *Elise G.* is small and primitive compared to many of the boats fishing Georges Bank today. Large European trawlers have for many years fished off the U.S. coast. These vessels, especially those from West Germany and the Soviet countries, are many times larger than the *Elise G.*, carry crews of seventy men, and perform the entire job of catching, cutting, and freezing their catch on board. These trawlers are served by still larger factory and supply ships which provide provisions, repair facilities, and replacement crews. Many of the trawlers remain at sea for years at a time while their catch and replacement crews are regularly shuttled back to Europe.

The foreign ships take huge quantities of fish in their nets—50 tons at a time is not unusual. These fish are filleted by machine and immediately frozen into huge blocks and slabs of solid flesh. The blocks are kept frozen in the ship and are later shipped back to Europe. There, they are landed and, still frozen, sawn into smaller portions to be packaged and marketed. Most of the frozen fish sold in the United States is first caught by foreign vessels off the East Coast and landed in Europe. It is then shipped back across the Atlantic to the United States for sale.

During the 1960s the number of foreign vessels fishing on Georges Bank grew alarmingly. They were catching everything in sight. Their massive nets, the American fishermen said, were gouging the sea bottom so badly that great areas of Georges Bank were becoming lifeless deserts. Their tow wires and rigging were so strong that when the foreign ships snagged wrecks and boulders they often dragged them for miles across the bottom before they finally got free. Charts of snags became useless to fishermen who had them because the foreign ships were all the time mixing things up.

The foreign fleet went after haddock and in a year or two the haddock were virtually wiped out. They went after yellowtail and as quickly decimated them.

"Stop the foreigners," the American fishermen pleaded. "Everything is being destroyed." But the U.S. government claimed authority to a distance of only 12 miles off its coast. "There is nothing we can do," the government replied. "But if we don't do it, no one will. We must," replied the fishermen, and the fisheries biologists agreed.

By the most ancient of legal traditions, a fish in the sea belongs to whomever is able to catch it. In the past, when there were many more fish than men to eat them, this tradition seemed proper. Fish were free wealth, so first come, first served. But recently everything has changed. There are not enough fish to go around. If an unlimited number of vessels are permitted to go after what fish there are, all will be caught and everyone will be the poorer.

Finally, it became obvious that something had to be done to save what was left of the U.S. fishing grounds. In 1977, when many feared that it was already too late, the United States and Canada extended their jurisdiction to 200 miles offshore. With the 200-mile limit came the means of regulating the fishery.

On the basis of survey work done by the National Marine Fisheries Service, it was first estimated how fast each fish species could be expected to reproduce itself and what the maximum population of each species ought to be under ideal conditions. Catch quotas were established that would allow the numbers of each species to build back up to its ideal size. After this, the quotas are adjusted so that each year only that number of fish are taken that can be replaced by the next year's spawning. This annual quota, set to maintain a high, stable population, is called the "maximum sustained yield" for that species of fish.

Each foreign country wishing to take fish within the 200-mile limit is issued a permit allowing it to take a certain percentage of the annual quota. Vessels are inspected at sea by Coast Guard and Fisheries Services personnel to insure that they are honoring their quotas. Although the system is designed to give American fishermen a healthy share of the total catch, it is not intended that foreign vessels be excluded altogether. The system is intended to protect the resource for everyone.

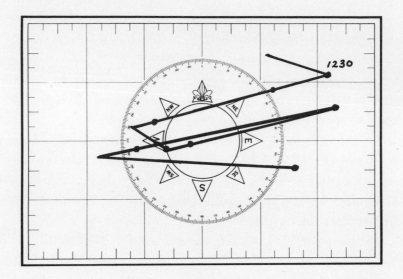

1230

Kaare signals to haul back for the third time. The net contains twenty-two baskets of yellowtail and one of cod. The *Elise G.* swings around in the opposite direction and the net goes right back into the water.

The men are now about midway through the last day of the trip. They have been working tremendously hard, eighteen to twenty hours each day since Thursday. Tom, Rob, and John are now entirely silent. Each is nearing the limit of his endurance. The men realize that if they were to slip while carrying the heavy baskets of fish they would not be able to catch their balance in time to save themselves from falling. During the first part of the trip each man would often carry, by himself, a large basket, weighing 100 pounds when full. Now they wait for help or else drag the basket across the deck by one handle. More and more they depend on one another for assistance.

There is an unstated rule that if one man is working on deck, then everybody will work there until the job is done. A man will almost never work alone on deck. They all know that a man alone could be hurt or caught in a dangerous situation from which he might not be able to escape without help.

Another incentive to this absolute cooperation is the manner in which the men on the *Elise G.* are paid; each man gets exactly the same amount. If each is paid the same, including Kaare, then it is expected that everyone will do the same amount of work.

The amount that each man will receive at the end of each trip depends on how many fish they are able to catch and upon the price that they receive for those fish when they arrive in port. They work by a share system, known as "a broken 45" in which all of the money from the catch, known as the "gross stock," is divided up in the following way:

First, the expenses of the trip are deducted from the gross stock. These expenses include the cost of food, fuel, ice, and lumpers. For an eight-day trip to Southeast Part this will amount to perhaps $1,500.

Second, 45 percent of the remaining cash is subtracted to pay for the maintenance and mortgage of the vessel.

The rest of the cash, about 50 percent of the gross stock, is then evenly divided among the people on board.

There are several other systems by which each man's share might be figured. Under some systems the expenses are taken out *after* the 45 percent for the vessel is deducted. This means that the vessel and its owner will get more and the crew will get less. Under other systems the vessel may get more or less than 45 percent. On many vessels those members of the crew with specialized duties such as captain, engineer, cook, or mate will get a small fee for each trip in addition to their regular share.

Obviously, the number of crew among whom the stock is shared has a great deal to do with how much each man will get. This results in a constant pressure to work with as few crew as is humanly possible. By stopping for a few hours each night when the fishing is usually not so good anyway, the four men on the *Elise G.* get barely enough sleep to make it through the trip. If they were to go for longer trips or if they were to fish through the night as many other boats do, they would have to take on at least one more man. The crew feel that the extra work is worth the extra money, and so long as they are able to keep up with the exhausting pace set by their fellow crewmen, they make very good money and are happy.

But should a man hurt his hands or his back or his knees, or should he grow old, he will no longer be able to keep up. There are few fishermen who are able to work on deck beyond the age of fifty-five.

Fishermen believe that it is impossible to teach a man to work. Certainly, you can show him what to do, but whether or not he chooses to do it well is entirely up to him. On the other hand, they feel that even the most incompatible people ought to be able to get along together. On almost every boat there are stories of crewmen who thoroughly hate one another but who work side by side in perfect harmony for years. There are also plenty of stories of captains who have fired close friends, brothers, sons from the crew because, however loved they might be, they were not working their share.

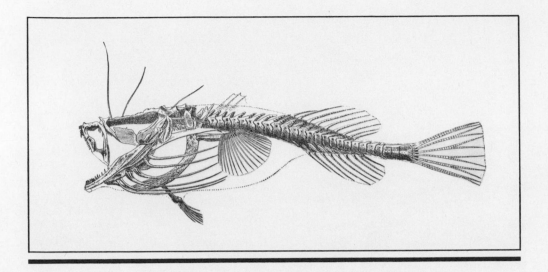

Angler

LOPHIUS PISCATORIUS

    Angler, headfish, goosefish, mollykite—all are names for a creature mostly jaw and tooth. Said to eat anything that moves, you will find in their stomachs flounder, haddock, cod, and many other things from the skeletons of sea birds to smaller members of their own kind. They are not only voracious, but cannibals as well!

    Anglerfish were once thrown back as ugly, inedible, and worthless. In the last few years the public has begun to develop a taste for them. This has given them a market value of about 30¢ a pound. They are not so ugly anymore.

The entire forepart of the fish is contained within the span of its jaw. The eyes, the brain, the gullet, and the intestine are all packed into the grotesque head. The skin is flabby brown. The eyeballs are speckled like shattered amber glass on black ice. The mouth is cavernous and rimmed with needle teeth. Like that of a bulldog the lower jaw protrudes.

The fish may reach a weight of 70 pounds, but lying flat on the sand, jaw closed, it looks for all the world like a mossy stone. From a point midway between its eyes springs a slender stalk at the end of which is a small tab of flesh. This, the ichthyologists assert, is a lure, a bait which when bobbed at the end of the stalk most resembles a small, helpless fish.

Other fish are attracted to the angler's bait by the promise of an easy snack. They lunge at the bait and in so doing seal their fate. For the angler in that instant gapes with a great suction, and at the clap of doom it seems the very floor of the ocean has broken to suck the unsuspecting in.

By making other fish greedy, the angler itself grows fat. But in a way it may fall victim to its own greed for by dangling its bait endlessly in front of its own face it may someday, if its jaw grow great enough, attempt to swallow itself.

In the stomach of a larger angler John finds the still intact body of a second angler almost as large. Perhaps the larger has lured and eaten its own little brother.

"Lookit," John calls over to Rob, "two heads in one."

"Keep the big one," says Rob, "We'll have it for dinner."

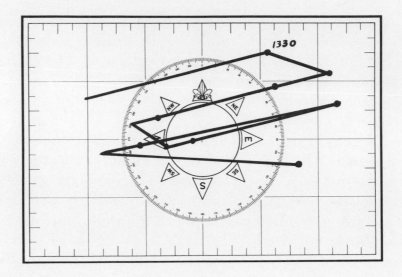

1330

Long before all of the fish from the last tow have been cleaned and put below, Kaare suddenly gives the signal to haul back. The engine is still roaring, but the *Elise G.* has stopped dead. The water at her stern is torn white by the beating of the propellor. The net has hung solidly on something. The winches are started and the *Elise G.* is dragged steadily backward toward the net, engine still churning. Finally the stern is directly over the net; the wires are straight up and down. Each winch is able to pull 15 tons. The strain sinks the stern deep in the water so that better than half the ramp is submerged. But, 250 feet below, the net does not budge.

Kaare reverses the engine so that the *Elise G.* backs over the net. The wires now point toward the bow. The winches continue to strain and slowly at first, then more easily, begin to wind in. The crew expect to see the net come up in shreds but, incredibly, it appears at the surface without a tear. It also contains ten baskets of yellowtail—a respectable haul for a tow that ought to have been disaster. The net is emptied and put right back in the water.

What hung up the net was probably a sand hill. In many places on Georges Bank the sand is heaped into mounds by the strong current much like sand dunes at the beach are formed by the wind. If the net runs squarely into one of these, it digs in and fills with sand. Once the net has been freed the sand pours through the mesh so that by the time it gets to the surface there is no evidence of what gripped the net minutes before.

Hanging-up can cause great damage. Sometimes nets are entirely lost if they become entangled in a wreck or a field of boulders. But there may also be danger to the ship itself.

Not long ago a vessel very much like the *Elise G.* was towing down wind in rough weather. As she moved along at 3 to 4 knots, the large seas rolled up from behind and passed easily under her. Then she hung up.

She stopped dead and the next sea, an especially large one, thundered in over the stern, swept the length of the work deck, and deluged through an open door into the engine room.

The engine drowned and the ship was entirely disabled, with no power to turn to face the seas, no power for the winches to slack out wire. Seas continued to pile in over the stern. The only hope of saving the ship was to cut the tow wires, but a moment's hestitation and the continued swamping of the deck under breaking seas quickly made this impossible. Next, the fishhold flooded, putting the entire work deck under water and leaving the vessel floating only on the watertight compartments beneath the living quarters. Realizing that their vessel was finished, the crew radioed for help. The Coast Guard arrived barely in time to get everyone off.

Sometimes the net may pick up a bagful of stones or tons of weed and thick mud. Another time it may get thousands of dogfish, or a peculiar orange sponge the men call "monkey dung." Sometimes so much trash is caught that the net is impossible to bring back aboard. To be emptied it must be slit open.

During her first winter of fishing off New Jersey, the *Elise G.* made a huge haul of porgies. In trying to bring the fish-choked net aboard, the gears of the net drum broke. The capstan couldn't raise the cod end an inch. They tried unsuccessfully to bring the net alongside to lift it bit by bit over the rail, they tried splitting, but nothing worked. The weather began to get worse. Finally there was nothing to be done but to steam full speed ahead until the net burst, letting the huge catch of fish drift away. Kaare estimates that the market value of those fish was $40,000, all of it lost.

Although nothing so dramatic is happening today, there is an increasing amount of empty clam shell coming up with each tow. The heaps of shell make it ever more difficult to pick out the valuable fish. They also add greatly to the work of shoveling the trash back over the side. The men are growing ever more tired as the work grows ever more difficult.

Also, they are becoming weak from hunger. The tremendous amount of energy they expend each day requires a huge quantity of food. Since noon Rob has been throwing off his gloves and ducking into the galley whenever he gets the chance; between jobs on deck he has been putting lunch together. When Kaare comes down from the

wheelhouse to examine the latest tow of fish, he sees that if the crew have to clean up the entire pile without first eating lunch they will reach a state of fatigue that will weaken them for the rest of the day. They have eaten nothing for eight hours.

"I guess we leave all this right here on deck and we go have something to eat," says Kaare.

The net is kept on the drum and the pile of fish is left as it is. The men strip off their oilskins, slap fresh water on their hands and faces, then sit down to the galley table.

There is a pot with a slab of corned beef boiled with an entire head of cabbage and a dozen potatoes. There is a pot of boiled peas. There are rolls and butter, milk, pickles and jam, mustard, and iced tea. All of the pots and dishes are jumbled on the table wherever they will fit. Each man takes a pot, serves himself, then passes it to someone else. There is a brief confusion of lids and crossed arms as each man piles his bowl with food. There is deep silence as each man eats.

Then Tom finds something else in his bowl. He fishes it out with his fork, holds it up, and asks Rob, "What's this?"

"It's a rutabaga."

"No, it's not. It's a turnip," offers John.

"All the same thing, I guess," says Kaare.

Tom replies, "I didn't know you ate turnip with beef and cabbage."

"This isn't beef and cabbage," says Rob.

"What is it, then," asks John.

"It's a 'New England boiled dinner.' "

"No, that has carrots in it, too," says John.

"Be thankful you got something, John," snaps Kaare.

Kaare has cleaned his plate. He refills his coffee cup and climbs back up to the wheelhouse. He begins to maneuver the *Elise G.* into position for the next tow. Tom, Rob, and John hurry to finish what is left on their plates, then rise, throw their dishes in the sink, and climb back into their oilskins. They reach the work deck at just about the time that Kaare signals to begin the set. In moments the net is back on the bottom. They are fishing again and ready to work for another eight to ten hours.

They will not dine again until midnight.

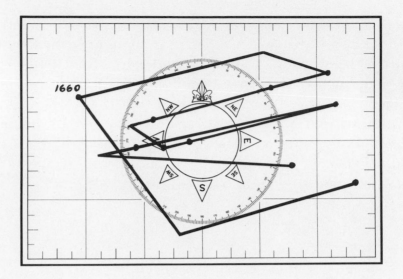

1600

While the net is being towed for the fifth time today, the crew take care of the pile of fish on deck. Shoveling the masses of shell over the side is back-breaking work. Kaare has a spare net on board the sweep of which is rigged with small rollers. The rollers should keep the mouth of the spare net a little off the bottom so that it will pick up less shell.

Kaare also has a hunch that farther east, in slightly deeper water, there will be more cod and haddock and fewer yellowtail. They are now almost to the limit of yellowtail while they are still well short of the limits on the other two fish.

The next time they haul back they have twelve baskets of large yellowtail, two of small, and one basket of cod. As soon as the fish are dumped out, the net they have been using is stripped off the drum and piled up at the side of the work deck. The spare net is then spooled onto the drum. Then, while the crew set to work on the pile of fish on deck, Kaare steams 3 miles to the southeast, into deeper water, then slows down and gives the signal to set. The roller net is put over.

Rob and John stand by the gallows ready to unhook the doors while Tom operates the winch. The wires stream out evenly. Then without a sound a connecting link in the wire jams against part of the door on the starboard side. The tow wire is held from running out. But the starboard winch continues to turn. Loose wire snakes across the deck behind Rob's back. He does not see the danger. For a few seconds Tom does not realize what is happening either. His vision is obstructed by the net they have piled up on the side of the deck. If it were the first day of the trip rather than the last, he would perhaps have realized what was happening sooner, but as it is, 100 feet of wire slither on the deck before he shouts, "LOOK OUT!" and immediately stops both winches. If the wire were to free itself at this moment it would whip like an ax. A man in its way would have no chance.

Rob turns and sees the danger. John at the same time shouts, "Over here!" and Rob dodges across the deck to take shelter behind the gallows legs on John's side. Tom gingerly reverses the winch and slowly retrieves the loose wire. Once everything is back under control he nods to Rob, who returns to his post. Tom's hands are trembling slightly. Otherwise he seems perfectly calm.

The rest of the tow proceeds without incident. When they haul back an hour later, they find less shell, fewer flounder, plus a good quantity of cod and haddock. Also, they find the belly of the net is again torn completely away.

It will be nearly dark before they will be able to set the net once more, and there will be time for only one more tow. The deep fatigue that was eased a little by lunch has come back with force. While cleaning up after the previous tow, Rob lost his balance while crossing the deck with a full basket. He fell on his hip and partially caught himself by twisting the basket underneath his body as he fell. He scrambled back to his feet, shaking his head, angry.

They are only a few hundred pounds short of the limit of yellowtail and a couple of thousand short of the less valuable cod and haddock. Why not call it quits and head back now, mend the net tomorrow while steaming?

Kaare firmly believes that the difference between a successful fisherman and an unsuccessful one is the determination not to quit.

"A man must not give up. If the trip is a bad one and you catch no fish, then you must stay out until you do catch fish. If you come home with nothing, then you must go out again. No, a man must not give up too soon."

Ever since lunch, Kaare has spent as much time on the work deck as he is able. He recognizes that the crew are played out. The work seems to double in difficulty with every tow. Yet these are the hours of the trip that make the difference between returning to port with a full boat or a partly full boat.

It is in these last hours that Kaare proves the absolute intensity with which he pursues his profession. He is all about the deck, lifting baskets, picking, cleaning, and sorting fish. He is relentlessly, constantly in motion.

"If you want to be the captain, then you got to be the first and also you got to be the last in everything. If you ask a man to work, then you must be sure that you are ready to work just as hard. No man will ever really work hard *for* another person. He will only work *with* him. So, if you are going to be the captain, well, you got to be the first and you got to be the last, too."

Kaare Gjertsen is not a tall man; he is 5′10″. Nor a heavy man; he weighs 146 pounds. But Kaare Gjertsen is the captain, and he is the first and the last aboard the *Elise G*.

Visitors

During the late afternoon two visitors arrive. One is a yellow-bird, female, tiny, and alert. The other is a young cowbird, brown and awkward. They both have flown a great distance. They are exhausted, hungry, and, with nothing but sea water around them, in a crisis of thirst.

Land birds are often encountered far at sea. The fishermen believe that they are lost after being blown far offshore by storms. This may be true, but it also may be true that the birds simply go astray in fog or at night. Very little is known about how birds tell where they are going even as some of them migrate from one end of the earth to the other.

Once they land on the *Elise G.*, the lost birds immediately begin to search for food and water. The cowbird paces up and down the work deck. It pecks at tiny crabs, mud shrimp, and seaweed, but they are too salty and strange for him to eat. He is quite fearless of the men. Just as if he were at home in a dairy pasture feasting on insects stirred up at the feet of grazing cattle, the cowbird scurries over the gear on deck while the men go doggedly about their work of harvesting fish. They barely notice the hungry brown bird underfoot.

In the meantime, the yellowbird flies off to the complicated rigging and antennae that fit on the roof of the wheelhouse. There in a corner of the roof it finds a puddle of fresh water left by the thick fog of the night before. First, it drinks, then begins to hunt.

Carefully it probes each corner, every crack and tiny seam in the superstructure. This yellowbird is used to hunting in woodlands for its flies and moths. Surely it will starve on this welded, painted, and steel-hard environment. But suddenly from a tiny recess at the base of the radar antenna the female drives a pale moth. The moth flies aft to escape, but the bird flies faster, snaps it out of the air, and relands. She snips off the wings and swallows the rest. In minutes she catches another moth hiding under a power cable. More are found in the most unexpected places, in a coil of rope, the corner of a window; one flies out of the foghorn. The little bird gathers a feast all afternoon by carefully inspecting every inch of the *Elise G.*'s rigging and deck.

Meanwhile, the cowbird continues to hang around the work deck. Perhaps the bird hopes the deck will suddenly sprout grass and grow cows to save him from starvation. After several hours of disappointment, he flies away from the *Elise G.* heading east, not toward land but even farther to sea, eventually to fall and to drown.

The yellowbird stays on board. She hunts through the day, perches in the rigging at night, drinks water that distills from the night fog, and two days later returns to her woodland when the *Elise G.* makes port.

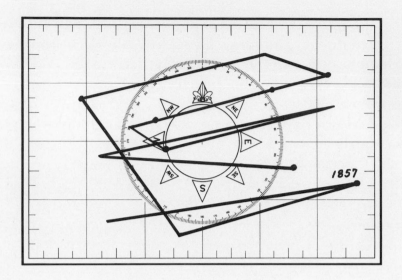

1857

Sunset. The gentle west wind has fallen away to a vast calm. It is warm. The air has gone soft and humid.

It is almost dark by the time the net is mended. It is reset, then the men turn once more to the endless job of picking, sorting, and cleaning fish. There are more cod than usual now. This means there is more gutting to do. They are still at it ninety minutes later when it is time to haul back for the last time.

"I think we got enough with this one," Kaare says as he inspects the pile of mixed fish that pours from the net. There are certainly enough yellowtail to complete the 25,000-pound total; there are also a lot of cod and a few haddock. An anglerfish thrashes and gapes like a reptile. Dogfish sweep their tails from side to side as if still in the water, eternally swimming. Sculpin bleat like tiny goats in the dusk.

And so, the last cleanup begins. Everyone, tired to the point of exhaustion, turns to. Kaare grabs a shovel and begins to pitch the trash overside. Tom, Rob, and John stalk through the pile with their picks. The deck lights are on. Everything is changed to a mercury green. The delicate coloration of the flounders is appalled in this unnatural light making it difficult for the pickers to tell one species from another. Often, they must hold a fish up to examine it before they can tell if it's a yellowtail, lemon sole, gray sole, or just a worthless four-eye or windowpane.

The flounders are picked into one pen; the cod and haddock into the other. Tom drops his pick to begin gutting fish. He sets his box by the rail, facing the sea. In a moment, Rob also puts down his pick and sets up another box next to Tom. John picks through the rest of the pile while Kaare continues with the shovel.

The *Elise G.* is stopped dead in the water. As soon as the night became fully dark the fog reappeared—first a haze, then a mist, and now a dungeon-thick cloud wreathes around the deck lights and blocks out everything. The universe has shrunk to a few feet of water and steel in the beam of the deck lights. The men are wearing their oilskin jackets now to protect against the chill fog. They move stiffly and with a slow determination in their bulky clothing.

The trash tumbles over the side. Some of the fish float, everything else sinks. The water is fantastically clear. Because of the deck lights, white objects such as the undersides of flounder can be seen fathoms down as they fall toward the bottom.

Sharks appear out of nowhere. In a matter of minutes there are five of them, then eight, then still more. They are what the fishermen call blue sharks because their backs glow with that electric color as they twist in the watery light.

When they take a fish underwater, they roll on their sides at the instant they strike. The color in their backs flashes as they glide in slow orbits around and through the pool of light beneath the ship.

When they take a fish on the surface, they approach straight ahead and lift their heads half out of water. They swim right over the floating fish and swallow it down with a smacking of gums and, sometimes, a slap of the tail. The sharks together perform a concert of splashing around the *Elise G.*

And sea birds gather. Few shearwaters come after dark, but dozens of petrel flutter around the ship. They hover about, in and out of the light, dipping and dipping at the sea surface. These little birds are called Wilson's Storm Petrel, *Oceanites oceanicus.* They are called by the fishermen "Mother Carey's hens," or "Carey chickens" for short. In superstition they are believed to be the ghosts of drowned sailors. In science they are thought to be the most abundant species of bird on earth.

Carey chickens nest in the far South Atlantic, on the very coast of Antarctica. During what is the South Atlantic winter (our summer) they migrate north. Because no one knew until quite recently where these birds nested, it was long believed that they laid their eggs in the warm waters of the Gulf Stream and that the young flew full fledged from the warm, brooding sea. It was thought that these birds might never touch land.

One petrel brushes the rail of the ship with its wings and falls onto the deck. In the confined space it is unable to take off. Tom